Elgin Community College Library
Elgin, IL 60123

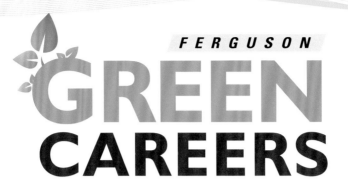

ENERGY

BOOKS IN THE GREEN CAREERS SERIES

Business and Construction
Communication, Education, and Travel
Energy
Environment and Natural Resources
Law, Government, and Public Safety
Science

FERGUSON
GREEN CAREERS

ENERGY

PAMELA FEHL

ELGIN COMMUNITY COLLEGE
LIBRARY

Ferguson Publishing
An imprint of Infobase Publishing

Green Careers: Energy

Copyright © 2010 by Infobase Publishing

All rights reserved. No part of this book may be reproduced or utilized in any form or by any means, electronic or mechanical, including photocopying, recording, or by any information storage or retrieval systems, without permission in writing from the publisher. For information, contact:

Ferguson
An imprint of Infobase Publishing
132 West 31st Street
New York NY 10001

Library of Congress Cataloging-in-Publication Data
Fehl, Pamela.
  Energy / Pamela Fehl.
    p. cm. — (Green careers)
  Includes index.
  ISBN-13: 978-0-8160-8150-9 (hardcover : alk. paper)
  ISBN-10: 0-8160-8150-6 (hardcover : alk. paper) 1. Renewable energy sources—Vocational guidance. 2. Green technology—Vocational guidance. 3. Energy industries—Vocational guidance. I. Title.
  TJ808.F445 2010
  621.042023—dc22   2009043353

Ferguson books are available at special discounts when purchased in bulk quantities for businesses, associations, institutions, or sales promotions. Please call our Special Sales Department in New York at (212) 967-8800 or (800) 322-8755.

You can find Ferguson on the World Wide Web at http://www.fergpubco.com

Text design by Annie O'Donnell
Composition by EJB Publishing Services
Cover printed by Bang Printing, Brainerd, MN
Book printed and bound by Bang Printing, Brainerd, MN
Date printed: April 2010
Printed in the United States of America

10 9 8 7 6 5 4 3 2 1

This book is printed on acid-free paper.

All links and Web addresses were checked and verified to be correct at the time of publication. Because of the dynamic nature of the Web, some addresses and links may have changed since publication and may no longer be valid.

Contents

Introduction	vii
Bioenergy/Biofuels Workers	1
Coal Gasification Engineers	13
Electrical Engineers	22
Energy Conservation Technicians	35
Geotechnical Engineers	46
Green Vehicle Designers	54
Hydroelectric Engineers	65
Nuclear Engineers	76
Nuclear Reactor Operators and Technicians	86
Petroleum Engineers	98
Petroleum Technicians	109
Power Plant Workers	122
Renewable Energy Workers	130
Solar Engineers	148
Wind Power Engineers	160
Further Reading	169
Index	171

Introduction

Coal, oil, and natural gas are nonrenewable energy sources that have an expiration date. Supplies can only last for so long, and when they run out, what will replace them? Green energy is a growing industry because of the increased need for alternative, clean energy sources that have less impact on the environment than fossil fuels, and that are accessible and—most importantly—replenishable.

Using renewable energy is not a new practice. As you well know, early man used wood to create fire to cook food and heat homes. And wind, water, and solar power were originally used to power boats as well as heat and cool buildings. New and emerging technologies are helping today's inventors, scientists, researchers, engineers, and others to take bioenergy and biofuels to the next level. Many are exploring and testing methods to produce cleaner energy and reduce carbon emissions.

Worldwide awareness of environmental issues has fueled the public's interest in using alternative energy sources. More people are seeking ways to reduce their energy bills while also doing something good for the planet. Companies are converting their manufacturing and production processes to comply with stricter federal, state, and local environmental laws and regulations. Funding for energy research and energy-conversion projects is on the rise. Overall, this is an exciting time to work in the energy industry and be involved in the green-collar industry in general. According to a joint study of the green economy by the American Solar Energy Society (ASES) and Management Information Services, in 2007 the renewable energy industry grew three times as fast as the U.S. economy, with solar thermal, photovoltaic, biodiesel, and ethanol sectors at the lead, each with more than 25 percent annual revenue growth. In that same year, more than 9 million jobs in the United States were in renewable energy and energy efficiency (RE&EE), representing $1,045 billion in revenue—an increase over the 8.5 million jobs and $975 billion in RE&EE revenue in 2006. And renewable energy job growth is expected to continue: ASES forecasts that by 2030 there may be as many as 37 million jobs in the U.S. RE&EE industries.

The jobs highlighted in the *Energy* volume of the Green Careers series are by no means the only ones that exist in the field, but they will give you an idea of some of the areas in which you can specialize. Featured job profiles include bioenergy/biofuels workers, coal gasification engineers, electrical engineers, energy conservation technicians, geotechnical engineers, green vehicle designers, hydro-electric engineers, nuclear engineers, nuclear reactor operators and technicians, petroleum engineers, petroleum technicians, power plant workers, renewable energy workers, solar engineers, and wind power engineers.

To give you a full, yet brief, introduction to the work, and to help you learn if your interests and skills match the requirements, we've broken each profile into 12 sections:

- **Quick Facts** is a fast summary of the basic information about the job, including salary range and a quick look into the future.
- The **Overview** is a snapshot—job responsibilities are spelled out in several sentences.
- Some green energy jobs have been around for a long time, while others are fairly new and evolving. They all got their start somewhere, though, and the **History** section tells you how and why they began.
- **The Job** digs deeper into the work, spelling out the day-to-day duties. You will see that some profiles also include comments and insights from people working in the field.
- The **Requirements** section can help you plan ahead. It recommends course work for high school, undergraduate, and, if needed, postgraduate studies. And **Other Requirements** explains the character traits and additional skills needed to enjoy and thrive in this work.
- Suggestions for ways to learn more about the job and industry are found in the **Exploring** section. Here you'll see book and magazine recommendations, Web site and online video referrals, and more.
- **Employers** focuses on types of industries and companies that hire the worker that's featured and may include statistics regarding the number of professionals employed in the United States, and the states and/or cities in which most professionals are concentrated. Statistics are often derived from the U.S. Department of Labor (DoL), the National Association of Colleges and Employers, and professional industry-related associations.
- **Starting Out** gives you tips on the steps you can start taking toward learning more about this job and getting your foot in the door.

- The **Advancement** section sheds light on the different jobs and specialties professionals can move up to in the green energy field.
- **Earnings** gives you salary ranges for the specific job and closely related jobs. Information is based on surveys conducted by the DoL, and sometimes from such employment specialists as Salary.com.
- The **Work Environment** section describes the typical surroundings and conditions of employment—whether indoors or outdoors, noisy or quiet, social or independent. Also discussed are typical hours worked, any seasonal fluctuations, and the stresses and strains of the job.
- The **Outlook** section tells you the forecast for the job. Does it have a bright future, or is it a risky business? Most jobs depend on the economy. When things are up, jobs abound. When things slow down, fewer jobs exist and competition heats up. The forecast may be based on DoL surveys, professional associations' studies, or experts' insights on the field.
- Each profile ends with **For More Information**, providing you with listings and contact information for professional associations you may want to join, and other resources you can use to learn more about the job.

Bioenergy/Biofuels Workers

 QUICK FACTS

School Subjects
Math/science
Technical/shop

Personal Skills
Communication/ideas
Helping/teaching
Technical/scientific

Work Environment
Indoors and outdoors, depending on job
One or more locations

Minimum Education Level
Bachelor's degree

Salary Range
$18,030 to $59,750 to $145,920+

Certification or Licensing
None available

Outlook
As fast as, or much faster than, the average

OVERVIEW

Energy that is derived from biomass (organic material such as wood, plants, or animal wastes) is known as *bioenergy*. Bioenergy can be used to generate electricity and produce heat. It can also be used to produce *biofuels*, which are used in place of fossil fuels to power vehicles and for small heating applications. A wide range of jobs are available in the bioenergy and biofuels industry for people with various backgrounds and interests, including scientists, biologists, researchers and research technicians, plant managers, product managers, and sales engineers.

HISTORY

Renewable energy sources have been around for thousands of years. Wood, corn, soy, aquatic plants, and animal waste are a small sampling of the types of organic, replenishable materials that can be converted into fuels to power cars and trucks and provide heat and electricity to homes and buildings. In fact, wood is one of the earliest biofuels—cavemen were the first to figure out how to use it to improve their quality of life. With fire, food was cooked and hot, homes were warm, and nights were brighter. Vegetable oil is another early biofuel. In 1990 Rudolf Diesel, the inventor of the diesel-fueled engine (as you may have guessed), demonstrated his engine at the World Exhibition in Paris, France, by using peanut oil to power it. Ford's Model T, which was produced from 1903 to 1926, was designed to run on hemp-derived biofuel. Back then, the abundance (not to mention the lower price and better efficiency) of fossil fuels pushed interest in biofuels to the wayside. Petroleum, among others, became the star.

As history has proven, one side effect of fuel shortages and energy crises is usually renewed interest in alternative energy sources. When faced with acute fuel shortages during World War II, Germany turned to potato-derived ethanol and wood-derived methanol as sources of fuel. The fuel crisis in the 1970s highlighted America's dependence on foreign-imported oil and the need to explore other sources of energy. America's consumption of foreign oil was at a high by the early 1970s, while production of oil on U.S. lands was at an all-time low. The fuel crisis started in 1973, when Middle Eastern countries, to illustrate their anger over outsiders' involvement in Arab-Israeli conflicts, placed an embargo (meaning a legal stoppage) on petroleum exports to Western Nations, including the United Stated and the Netherlands. As a result, awareness of natural resources and energy conservation grew. More people started buying smaller, fuel-efficient cars, as opposed to the gas-guzzlers that had once been popular. Carpooling and mass transit were heavily promoted as ways to save energy and money. And interest in and funding for bioenergy and biofuels research grew.

THE JOB

Bioenergy can be derived from wood, construction and consumer waste, landfill gas, and liquid biofuels such as ethanol for use in generating electricity, producing heat, and fueling vehicles. The United States gets approximately 4 percent of its energy from bioenergy, including wood and crops such as corn and soybeans.

There are a variety of jobs in the bioenergy and biofuels industry, from scientists and researchers to engineers, construction workers, product and plant managers, and a whole host of others. People are needed to create and improve the products and the technology,

A technician takes a sample of ethanol at a corn ethanol plant in Shenandoah, Iowa. Corn ethanol is a biofuel that is primarily used in the United States as an alternative to gasoline and petroleum. *David Nunuk/Photo Researchers, Inc.*

oversee and manage the operations and staff, as well as to build the facilities in which the work is done.

Scientists and *biologists* may work in the areas of research and development, to help advance the ways in which bioenergy and biofuels are produced. For instance, a *bioenergy plant scientist* may be involved in bioenergy crop research, conducting studies on plant growth and development, and plant adaptation to environmental stresses. This scientist's work may involve molecular studies of plants, in an effort to understand gene compositions and functions. A *seed production scientist* helps organize and manage seed production programs for companies. As described in one Internet advertisement for employment with a biofuels company in the agribusiness sector, the job can entail researching and developing seed production techniques for bioenergy grass crops, and include selecting production sites, field design, seed crop establishment, flowering and pollination control, seed harvest and handling, and developing quality-control programs.

Biological technicians work closely with biologists to research and study living organisms. They work as laboratory assistants, helping to set up, operate, and maintain laboratory equipment. They also monitor experiments, make observations, calculate and record results, and develop research conclusions. They may work in biotechnology, conducting basic research on gene splicing and recombinant DNA and applying knowledge and techniques to product development for biofuels and bioenergy. *Agricultural technicians* have similar roles in providing laboratory and research assistance, with their subject matter being crop production and processes. They conduct tests and experiments to improve the yield and quality of crops, and to help increase plants' resistance to disease, insects, or other hazards.

A *biofuels product manager* will work closely with business development managers, business analysts, and other product managers to help strategize business plans, product development, and product launches of biofuels. One posting for an ethanol product manager included job responsibilities such as interpreting customer and market needs and translating this information to research and development groups. Problem-solving abilities, strong communication skills, and the knack for analyzing data and communicating and presenting it clearly to different groups of people are required to succeed as a product manager. Other requirements may include guiding and participating in conferences, helping to create product strategies (including functionality, pricing, etc.), product application development and prototype testing, and handling communications with external groups.

Biofuels plant managers oversee all operations of biofuels and biorefinery plants; they are responsible for everything from machinery functionality to staff performance. A recent posting for a biofuels plant manager in Minnesota included these job responsibilities:

- Ensure production is efficient and maximized
- Provide direction to ensure proper levels of maintenance as well as compliance with safety and environmental regulations
- Direct continuous improvements and quality-assurance efforts
- Lead cross-functional efforts to facilitate best practices and process initiatives
- Provide solid leadership; serve as a role model for the plant's management team
- Revise policies and procedures as necessary to achieve the highest levels of morale and working relationships
- Prepare and manage plant-wide budgets

People applying for this job also needed to have strong knowledge of chemical distillation, fermentation, and grain refining, processing, and extracting processes. They also needed prior experience in a petroleum, ethanol, biodiesel, biofuels, or chemical plant.

Engineers and *construction managers and workers* are also needed to create bioenergy and biofuels plants. *Civil, electrical, industrial,* and *mechanical engineers* develop designs for plants and process equipment using computer-aided design and computer-aided industrial design software. They work closely with architects, developers, business owners, construction crews, and others to make sure the work is done according to specifications. Construction managers coordinate the construction process, selecting and managing construction workers, and overseeing projects from the development phase to final construction. They may work as project manager, site manager, construction superintendent, project engineer, program manager, or general contractor.

REQUIREMENTS
High School
Course work in math, science, physics, history, English, and computer software programs will provide a well-rounded basis for this career. Foreign language classes are also useful. If your school offers environmental studies classes, take these as well. Some bioenergy

and biofuels jobs may require knowledge of machines, so it may be advantageous to take electronics, mechanics, and shop classes.

Postsecondary Training

Undergraduate requirements will vary depending on the job. Many companies and universities prefer to hire scientists, biologists, and researchers that have a bachelor's degree in science, and a Ph.D. in their specialty, which could be plant biology, biochemistry, physiology or genetics, to name only a few. Previous related research and project work may also be required for more advanced positions. Engineers may have a bachelor's or advanced degree in electrical, electronics, industrial, mechanical, or even civil engineering. Plant managers and product managers may have a bachelor's degree in business administration, management, industrial technology, or industrial engineering. Some jobs require a master's or Ph.D. in business, marketing, chemistry, biotechnology, or related fields, with experience in the biofuels industry. Some companies may hire plant or product managers with a liberal arts degree who match all other requirements of the job and can be trained while on the job.

Other Requirements

Most bioenergy and biofuels jobs require strong oral and written communication skills to write reports, present materials, and manage staff. Scientists, researchers, and engineers usually work on teams, so the ability to share information and deal with different people is essential to succeeding in this type of work. Some positions may require knowledge of computer modeling, digital mapping, global positioning systems (GPS), and geographic information systems (GIS). Plant managers and product managers need strong organization skills in their work, as well as leadership and management abilities. Engineers will need to be well-versed in computer-aided design (CAD) and computer-aided industrial design (CAID) software systems. Knowledge of foreign languages can also be extremely beneficial in this field.

EXPLORING

Keep up with developments and trends in bioenergy and biofuels by reading magazines and books. Pick an area of bioenergy and biofuels that interests you, then do an Internet search to find the companies that specialize in this area. If any are located near you, see if they have part-time or summer job openings or volunteer opportunities.

Learn More About It

Goettemoeller, Jeffrey and Adrian Goettemoeller. *Sustainable Ethanol: Biofuels, Biorefineries, Cellulosic Biomass, Flex-fuel Vehicles, and Sustainable Farming for Energy Independence.* Maryville, Mo.: Prairie Oak Publishing, 2007.

Mousdale, David. *Biofuels: Biotechnology, Chemistry, and Sustainable Development.* Boca Raton, Fla.: CRC Press, 2008.

Rosillo-Calle, Frank, Peter de Groot, Sarah Hemstock, Jeremy Woods, eds. *The Biomass Assessment Handbook: Bioenergy for a Sustainable Future.* London, England: Earthscan Publications Ltd., 2008.

Soetaert, Wim and Erik Vandamme, eds. *Biofuels.* New York, N.Y.: Wiley, 2009.

Wall, Judy. *Bioenergy.* Washington, D.C.: ASM Press, 2008.

You can also find resources, event listings, and job postings on the *Biofuels Digest* Web site (http://biofuelsdigest.com). The site also features a useful article called "The Hottest 50 Companies in Bioenergy (2008–09)," with direct links to each of the companies.

EMPLOYERS

Scientists, biologists, and researchers work for universities, laboratories, and research institutes, as well as government agencies and private firms. According to the U.S. Department of Labor (DoL), there were about 92,000 environmental scientists (and hydrologists) employed in the United States in 2006. About 35 percent of all environmental scientists worked for state and local governments. Another 21 percent worked for management, scientific, and technical consulting services; and 15 percent worked for architectural, engineering, and related services. About 8 percent worked in the federal government, and 2 percent were self-employed.

The DoL reports that about 87,000 biological scientists were employed in the United States in 2006. (This number does not include the many who held biology faculty positions in colleges and universities, as these were categorized as postsecondary teaching jobs.) Federal, state, and local governments employed about

39 percent of all biological scientists. The U.S. Departments of Agriculture, Interior, and Defense and the National Institutes of Health were the main employers of federal biological scientists. The remainder worked in scientific research and testing laboratories, the pharmaceutical and medicine manufacturing industry, or colleges and universities.

Agricultural scientists held about 33,000 positions in 2006. They worked for federal, state, and local agencies; agricultural service companies; commercial research and development laboratories; seed companies; wholesale distributors; and food products companies. About 16 percent were self-employed as consultants.

Science technicians (including agricultural and biological technicians) held about 267,000 jobs in 2006. Government agencies, and scientific and technical service firms were their main employers.

Engineers held approximately 1.5 million jobs in the United States in 2006. About half worked in manufacturing industries and in the professional, scientific, and technical services sector. And of the 157,000 industrial production managers employed in 2006, about 80 percent (125,600) worked in manufacturing industries.

STARTING OUT

Internships, part-time jobs, and volunteer gigs are great ways to explore the bioenergy and biofuels field and see if this type of work suits you. Visit the Web sites of professional associations in the areas that interest you and see if there are opportunities to volunteer and get involved. Some resources are listed at the end of this profile to help you start exploring the field. You can also find useful information about the renewable and clean energy industry by visiting the U.S. Department of Energy's Clean Energy Jobs section (http://www1.eere.energy.gov/education/clean_energy_jobs.html).

ADVANCEMENT

Scientists and biologists can advance to more senior-management positions, such as department directors or regional supervisors. Plant and product managers with years of experience and proven track records can advance to regional manager, vice president, or a similar position of higher authority. Some workers may decide to freelance as consultants or start their own companies. Others can advance by teaching in colleges and universities, speaking at conferences held by industry associations, and writing papers and books on their area of specialty. Obtaining certification may not be required for some positions, but it can be attractive to prospective employers as it demonstrates knowledge in that subject. Getting a master's degree or Ph.D. can also help boost a candidate's chances for employment in certain areas of the bioenergy and biofuels field.

EARNINGS

According to the U.S. Department of Labor, in 2008 environmental scientists had median annual incomes of $59,750, with the lowest 10 percent earning $36,310 and the top 10 percent earning $102,610 or more.

Biological scientists' salaries in 2008 ranged from $35,620 to $101,030. Annual salaries for soil and plant scientists in 2008 ranged from $34,260 to $105,340. Biological technicians had median incomes of $38,400, with the lowest annual salary of $24,530 and the highest of $62,260 or more. Agricultural technicians averaged $22,190 to $53,880 per year.

Industrial production managers earned median annual incomes of $83,290 in 2008, with the bottom 10 percent bringing home $50,330 or less, and the top 10 percent earning $140,530 or more. In 2008, construction managers had annual salaries ranging from $47,000 to $145,920, with median incomes at $79,860. Construction laborers' salaries ranged from $18,030 for the bottom 10 percent, $28,520 for the middle 50 percent, and $54,030 or higher for the top 10 percent.

Engineers' salaries varied, depending on their specialty. For example, in 2008 environmental engineers earned annual incomes ranging from $45,310 to $115,430 or more; agricultural engineers brought home $43,150 to $108,470 each year; and chemical engineers earned $53,730 to $130,240 or more annually.

WORK ENVIRONMENT

Scientists, biologists, and research associates and technicians work indoors in laboratories and offices. They may spend some time outdoors conducting research on plants and crops and collecting samples. Engineers may also work in laboratories and offices, as well as outdoors, depending on the project. Construction managers and workers work on-site at construction sites and in offices. Product and plant managers work indoors in manufacturing facilities and plants, as well as in offices.

Work hours will vary, but most bioenergy and biofuels positions will require at least 40-hour workweeks, with additional hours occasionally needed on evenings and weekends to meet project deadlines. Some positions will require travel for research, meetings, and conferences.

OUTLOOK

The U.S. economic stimulus plan includes $500 million for "leading-edge" biofuels projects. With the government's interest in funding bioenergy and biofuels research and increasing the supply of biofuels, environmental science and research jobs in this field are expected to be on the rise in the years to come.

The U.S. Department of Labor predicts employment growth through 2016 for environmental scientists to be much faster than average, with private sector consulting firms offering the most job opportunities. Stricter environmental laws and regulations will increase the need for scientists and researchers working in the bioenergy and biofuels arena.

Biological scientists will also have decent employment opportunities through 2016 because of the focus on cleaning up the environment and reducing carbon and greenhouse gas emissions. According to the DoL, demand for biological scientists will continue to grow due to the need to study the impact of industry and government actions on the environment, and the need to develop ways to prevent and correct environmental problems. Environmental regulatory agencies will need biological scientists to help advise lawmakers and lawyers on environmental legislation. The need for alternative fuel will also increase the demand for biological scientists who specialize in biotechnology (e.g., using ethanol for transportation fuel).

Agricultural and chemical engineers will have employment growth as fast as average in the next few years. They will be needed to research and develop biofuels and biotechnology, and to create more efficient crops dedicated to biofuels production and renewable energy sources. Construction managers are also expected to have excellent job opportunities through 2016, according to the DoL. Those with bachelor's degrees, or higher, in construction science, construction management, or civil engineering, along with practical, related work experience will have the advantage in the job market.

FOR MORE INFORMATION

Agricultural Sciences
To learn more about agricultural science careers, visit

American Society of Agronomy
Crop Science Society of America
Soil Science Society of America
677 South Segoe Road
Madison, WI 53711-1086
Tel: 608-273-8080
Email: headquarters@agronomy.org
http://www.agronomy.org

Biological Sciences
For publications, networking opportunities, policy information, and other information related to biological sciences, visit

American Institute of Biological Sciences
1444 I Street, NW, Suite 200
Washington, DC 20005-6535
Tel: 202-628-1500
http://www.aibs.org

Chemical Sciences
For chemical scientist and chemical technician information, visit

American Chemical Society
1155 16th Street, NW
Washington, DC 20046-4839
Tel: 800-227-5558
Email: help@acs.org
http://www.acs.org

Construction
For more information about construction careers, visit

American Institute of Constructors
PO Box 26334
Alexandria, VA 22314-6334
Tel: 703-683-4999
Email: admin@aicnet.org
http://www.aicnet.org

Engineering

For more information about engineering schools and career resources, visit ASEE's Web site.

American Society for Engineering Education (ASEE)
1818 N Street, NW, Suite 600
Washington, DC 20036-2476
http://www.asee.org

Find out more about engineering careers by visiting

Junior Engineering Technical Society
1420 King Street, Suite 405
Alexandria, VA 22314-2750
Tel: 703-548-5387
http://www.jets.org

General Energy and Bioenergy Information

Learn more about energy issues and find out about upcoming events and conferences by visiting

American Council for an Energy-Efficient Economy
529 14th Street, NW, Suite 600
Washington, DC 20045-1000
Tel: 202-507-4000
Email: ace3info@aceee.org
http://www.aceee.org

Learn more about bioenergy and bioenergy programs by visiting

U.S. Department of Energy
http://www.energy.gov/energysources/bioenergy.htm

Product and Plant Management

To learn more about industrial production management careers, contact

Association for Operations Management
8430 West Bryn Mawr Avenue, Suite 1000
Chicago, IL 60631-3417
Tel: 800-444-2742
http://www.apics.org

Coal Gasification Engineers

QUICK FACTS

School Subjects
Chemistry/earth science
Technical/shop

Personal Skills
Mechanical/manipulative
Technical/scientific

Work Environment
Indoors and outdoors
One or more locations

Minimum Education Level
Bachelor's degree

Salary Range
$47,900 to $80,000 to $130,240+

Certification or Licensing
Required for certain positions

Outlook
About as fast as the average

OVERVIEW

Coal gasification engineers use engineering techniques and analysis to design coal gasification systems for chemical processes and power generation. They work closely with various engineers to help monitor and maintain coal gasification plants. They may work closely with vendors; travel to meet with suppliers and clients; and inspect and work on systems at construction sites.

HISTORY

Coal gasification is the process of converting coal into gaseous products. The conversion happens when coal or coal char (meaning residue of coal, as in charcoal) is reacted with a controlled amount of steam, oxygen, air, hydrogen, carbon dioxide, or a combination of these. The reaction creates a mixture known as syngas or synthesis gas. Whereas burning coal pollutes land, water, and air, and has contributed to global warming by increasing the carbon dioxide in the atmosphere, coal gasification produces "clean" gas. Once coal has been converted into a gas, pollutants (mercury, sulfur, particulates) can be removed and the gas can be used as oil to heat homes or fuel for cars. Coal gasification can be used to generate electricity in power plants; it can provide fuel for gas turbines, gas engines, or steam turbines, and can also be a fuel blend with natural gas and industrial gases.

In the 1800s coal gasification was used in the United States and England to produce gas for lighting and heat. By the mid-1800s more people were turning to natural gas as their fuel source, with many cities across the country using it to light street lamps. The first natural gas pipeline was drilled in Pennsylvania in 1859, kick-starting the spread of the industry. Natural gas was abundant and cheap and had replaced coal gasification as an energy source by the end of the 19th century. Coal gasification was not completely abandoned by all, however. When their petroleum supplies were cut off, the Germans made extensive use of coal gasification during World War II. And after the war, South Africa further developed goal gasification technology, to the point were it provided 50 percent of their gasoline and other fuels.

Underground coal gasification (UCG)—a less expensive process that causes less harm to the environment—was first written about and discussed in Germany and Russia in the late 1800s, but it wasn't until 1937 that the first commercial-scale UCG plant began operating (in Russia). The U.S. Bureau of Mines was involved in goal gasification research in the early 1900s. And U.S. and European scientists and researchers explored UCG in the late 1940s, but natural gas again replaced coal gasification as the fuel of choice.

It took the energy crisis of the 1970s—when oil prices skyrocketed—to spur the United States to revisit coal gasification as an alternative energy source. Since the 1980s, concerns about pollution caused by coal-burning processes have inspired more research and development of coal gasification as a way to generate clean power. In the mid-1980s, Southern California Edison's experimental Cool Water demonstration plant (near Barstow, California) was the first U.S. electric power plant to use coal gasification. The U.S. Department of Energy's Clean Coal Technology Program in the 1990s helped advance the goal gasification industry by providing federal cost-sharing for the first commercial-scale Integrated Gasification Combined Cycle (IGCC) power plants in the United States.

THE JOB

Rising fuel prices coupled with a greater supply of coal than oil in the United States are spurring more interest in coal gasification.

Did You Know?

- The United States produces about 20 percent (or 1.1 billion tons) of the world's coal supply—second only to China.
- Coal generates about 50 percent of the electricity used in the United States.
- As of 2007, the United States had approximately a 245-year supply of coal, providing it continued using coal at the same rate.
- Each person in the United States uses 3.8 tons of coal each year.
- U.S. coal deposits contain more energy than that of all the world's oil reserves.
- More than 2 million acres of mined land—an area larger than the state of Delaware—have been reclaimed since the late 1980s.
- Montana has the most coal reserves (119 billion tons). Wyoming produced the most coal in 2004 (400 million tons). And Texas is the top coal-consuming state, using about 100 million tons per year.

Source: American Coal Foundation, 2007.

With coal gasification, pollutants can be cleaned in the process, and old coal waste piles can actually be reclaimed and put to use. In the years to come, more large-scale commercial plants will be modifying their operations to produce eco-friendly fuels and meet carbon emissions standards, further exploring and developing coal gasification technology. One example of this exploration is the $800 million alternative energy plant being built by Emerson Process Management and Secure Energy Inc., in Decatur, Illinois—as of 2009 it is the first coal gasification-to-synthetic natural gas plant to be built in the United States in two decades.

An article in *Popular Mechanics*, published in 2007, explains the gasification process as follows:

1. In refinery plants, it's the "gasifier" that makes gasification happen. Carbon, water, and air are fed into the gasifier (which is really a big compartment where

"feedstock," the material to be used, is placed), and high pressure is applied to create "syngas"—a mixture of carbon monoxide and hydrogen. Rocks, dirt, and other noncarbon-based materials separate from the syngas and fall to the bottom of the gasifier. This waste is usually in the form of ashes or inert, glass-like materials that can be reused in concrete and road fill.

2. The next step involves getting the syngas out of the gasifier. At this point, the syngas is usually extremely hot and full of contaminants such as mercury, hydrogen sulfide, and ammonia. The syngas is cooled to room temperature by using various exhausts and filters, and the solid particles are removed.

3. To remove most of the mercury, the syngas is passed through a small bed of charcoal. The charcoal captures about 90 percent of the contaminant, and the contaminated charcoal is then disposed of at a hazardous landfill.

4. Acid gas removal units handle the last cleaning step, which involves removing sulfur impurities. These impurities are converted into sulfuric acid or elemental sulfur, each of which is a valuable byproduct.

5. A combustion turbine reheats the clean syngas. The turbine dilutes the syngas with nitrogen for control of NOx (a smog-making greenhouse gas) and burns it, driving a generator to make electricity.

6. A Heat Recovery Steam Generator (HRSG) recovers the leftover heat from combustion, which generates steam to power the internal turbine. Some of that air is compressed and can be channeled back to the air separation unit for oxygen, which is then reused within the gasifier.

7. To produce even more power, the steam generated in the HRSG is combined with the steam created in step 1 to drive a steam turbine. The steam then cools and condenses into water, which pumps back into the steam-generation cycle. In an IGCC plant, two-thirds of the total electricity produced comes from the gas turbine and one-third from the steam turbine.

Coal gasification engineers may work as *gasification plant production engineers.* In such capacity, their responsibilities may include working closely with operations and maintenance staff to track the performance of the gasification equipment. They may also work with the gas cleanup and system equipment vendors, and with performance engineers to maintain the cleanup system equipment and processes. They are familiar with silos, conveyors, metering, grinding, and drying equipment. Plant production engineers recommend equipment performance upgrades and modifications, evaluate spare parts inventories for the plant, develop and conduct tests to improve plant production capacity, and discuss and help resolve maintenance, operations, and technical issues. Most companies prefer to hire plant production engineers who have a master's degree in chemical engineering.

Some companies are setting up pilot projects and plants to produce cleanly, substitute natural gas from coal and renewable energy, recycle greenhouse gas releases, and create a new source of domestic transportation fuel. For example, Parsons, a California-based engineering and construction service company, plans to construct a 60-acre pilot plant in Holbrook, Arizona, for these environmentally concerned purposes. The aim of this research and development project is to "develop, test, and evaluate at an engineering scale, a fully integrated algae farm producing biofuels from recycled carbon dioxide emissions, a two-ton per day Hydrogasifier which produces SNG [synthetic natural gas], hydrogen from electrolysis, and fuel derived from algal lipids. The algae will be grown in saline water on non-arable land." The company was seeking *process engineers* with basic engineering knowledge, including mechanical engineering, to work on this project. A process engineer is involved in the design of coal gasification systems, and may also be involved in designing other systems such as electrolysis, algae farm, and lipid recovery, as is the case in Parsons' job requirements. A bachelor's degree in engineering is usually the minimum educational requirement for this job. Several years of work experience in goal gasification is preferred, and proficiency in computer-aided design (CAD) software and other engineering software programs is essential.

Engineers may also be required to interact with sales and commercial operations teams, make presentations to prospective clients, and field questions regarding technical aspects of the equipment and processes. *Project managers* and *directors* are responsible for hiring and overseeing staff, creating and managing work schedules, tracking work progress and creating reports, and liaising with various engineering teams as well as construction and mining crews and others. Those who own engineering consulting firms also handle the day-to-day tasks of running a business, including purchasing and maintaining office equipment and supplies; managing staff and performing staff reviews; marketing and promoting the business; negotiating contracts; and accounting and bookkeeping.

REQUIREMENTS
High School
Take classes in calculus, physics, chemistry, geology, English, computer science, and shop. Foreign language classes are also recommended. If your school offers environmental classes, take these as well. Advanced placement and honors classes will give you an advantage when applying to engineering schools.

Postsecondary Training
Most coal gasification engineers have a bachelor of science degree in engineering, with a specification in coal gasification, renewable energy, electrical, mechanical, or chemical engineering. They may also have degrees in petroleum or civil engineering. Course work includes math, statistics, physics, chemistry, geology, the study of fluid mechanics, material properties and behavior, thermodynamics, and design software programs specific to engineering, such as CAD and AutoCAD. Depending on the job, an advanced degree in engineering may be required.

Certification or Licensing
In the United States a professional engineer (PE) license is required for engineers who provide services to the public. To receive a PE designation, engineers must have a degree from an Accreditation Board for Engineering and Technology (ABET)-accredited engineering program, four years of related work experience, and pass a state examination.

Engineers who secure voluntary certification can advance to positions of greater responsibility and secure higher salaries. Organizations such as the American Academy of Environmental Engineers (http://www.aaee.net) and the Institute of Electrical and Electronics Engineers (http://www.ieeeusa.org) offer certification programs to professional engineers.

Other Requirements
To succeed in this job, engineers need to be detail oriented, able to juggle projects and meet deadlines, have strong verbal and written communication skills, and be capable of working with different teams of people daily. Depending on their specialty, engineers use

engineering software to produce and analyze numerous designs; to simulate and test machine, structure, or system operations; to generate specifications for parts; and to monitor product quality and control process efficiency. Solid knowledge of CAD and other engineering-related PC software is critical in this work.

EXPLORING

Learn more about alternative and renewable energy by reading magazines such as *Popular Science* and *Science.* You can also find information about coal gasification technology and research and development by visiting the Fossil Energy section of the U.S. Department of Energy's Web site (http://fossil.energy.gov/programs/power systems/gasification). If there's a coal gasification plant near you, call or email to see if you can set up a visit and tour.

EMPLOYERS

In 2006 more than 1.5 million engineers were employed in the United States, according to the U.S. Department of Labor. Manufacturing industries employed about 37 percent of all engineers, and another 28 percent worked in the professional, scientific, and technical services sector, primarily in architectural, engineering, and related services. Many engineers also worked in the construction, telecommunications, and wholesale trade industries. About 12 percent worked for federal, state, or local governments.

Coal gasification plants, power companies, engineering and construction groups, and consulting firms that provide renewable energy services hire engineers who specialize in coal gasification to manage and maintain plants; research, design, and improve equipment, technology, and systems; and, in some cases, to help develop and promote business. International travel or relocation may be required for certain jobs and projects.

STARTING OUT

Many coal gasification engineers get their foot in the door through apprenticeships and internships. Learn more about the industry and career opportunities by reading trade magazines and books and visiting the Web sites of such organizations as the American Coal Foundation (http://www.teachcoal.org) and the U.S. Department of Energy (http://www.energy.gov).

ADVANCEMENT

Advancement for coal gasification engineers can be from associate engineer to engineer, to senior engineer, project manager, or director. As more schools offer alternative energy engineering programs every year, engineers can expand their knowledge and technical skills by pursuing advanced degrees in specialty areas. They may also become partners in engineering firms, write books and papers on the subject of coal gasification, and teach at universities.

EARNINGS

Salaries vary for coal gasification engineers depending on their job responsibilities and level of experience. According to Indeed.com, in 2009 the annual income for coal gasification engineers ranged from $80,000 (for a coal gasification plant engineer) to $125,000 (for a chemical engineer in a coal gasification plant).

In 2008 chemical engineers had median annual incomes of $84,680, according to the U.S. Department of Labor. The bottom 10 percent took home about $53,730 per year, and the top 10 percent brought in upward of $130,240. Those working in architectural, engineering, and related services averaged $90,060 per year, and those providing scientific and research services had annual incomes of $95,380. Mechanical engineers had slightly lower annual incomes by comparison, starting at $47,900 and topping out at $114,740 per year.

WORK ENVIRONMENT

Coal gasification engineers work predominantly indoors in offices and in gasification and power plants, and also outdoors at construction and mining sites. Those that work in research and development may work in laboratories as well. Engineers work 40 or more hours per week. Travel may be required to visit plants, evaluate sites, attend conferences, and meet with prospective and current clients.

OUTLOOK

Employment growth for engineers in general is expected to be about as fast as the average for all occupations through 2016, according to the U.S. Department of Labor (DoL). While there are no forecasts specific to coal gasification engineering, the DoL predicts that chemical engineers will fare well in the next decade, with more work opportunities opening up in service-providing industries such as professional, scientific, and technical services, and particularly for research in energy. And while mining operations will continue to be restricted, research and development of coal gasification technology and processes will continue in an effort to combat rising fuel costs and tighter environmental regulations.

FOR MORE INFORMATION

For information about chemical engineering, visit these associations' Web sites:

American Chemical Society
Department of Career Services
1155 16th Street, NW
Washington, DC 20036-4839
Tel: 800-227-5558
http://www.chemistry.org

American Institute of Chemical Engineers
Three Park Avenue
New York, NY 10016-5991
Tel: 800-242-4363
http://www.aiche.org

For free student materials (booklets, brochures, posters, videos) about coal, electricity, and land reclamation issues, contact

American Coal Foundation
101 Constitution Avenue, NW, Suite 525 East
Washington, DC 20001-2133
Tel: 202-463-9785
Email: info@teachcoal.org
http://www.teachcoal.org

Learn more about coal and other fuels by visiting the Web site of the DoE.

U.S. Department of Energy (DoE)
1000 Independence Avenue, SW
Washington, DC 20585-0001
Tel: 202-586-5000
http://www.energy.gov

Electrical Engineers

 QUICK FACTS

School Subjects
Computer science
Mathematics
Physics

Personal Skills
Mechanical/manipulative
Technical/scientific

Work Environment
Primarily indoors
One location with some travel

Minimum Education Level
Bachelor's degree

Salary Range
$52,990 to $82,160 to $125,810+

Certification or Licensing
May be required

Outlook
Slower than the average

OVERVIEW

Electrical engineers apply their knowledge of the sciences to working with equipment that produces and distributes electricity, such as generators, transmission lines, and transformers. They also design, develop, and manufacture electric motors, electrical machinery, and ignition systems for automobiles, aircraft, and other engines. There are approximately 153,000 electrical engineers employed in the United States.

HISTORY

Electrical engineering had its true beginnings in the 19th century. In 1800 Alexander Volta made a discovery that opened a door to the science of electricity—he found that electric current could be harnessed and made to flow. By the mid-1800s the basic rules of electricity were established, and the first practical applications appeared. At that time, Michael Faraday discovered the phenomenon of electromagnetic induction. Further discoveries followed. In 1837 Samuel Morse invented the telegraph; in 1876 Alexander Graham Bell invented the telephone; Thomas Edison invented the incandescent lamp (which we know as the light bulb) in 1878; and

Nicholas Tesla invented the first electric motor in 1888 (Faraday had built a primitive model of one in 1821). These inventions required the further generation and harnessing of electricity, so efforts were concentrated on developing ways to produce more and more power and to create better equipment, such as motors and transformers.

Edison's invention led to a dependence on electricity for lighting our homes, work areas, and streets. He later created the phonograph and other electrical instruments, leading to the establishment of his General Electric Company. One of today's major telephone companies also had its beginnings during this time. Alexander Bell's invention led to the establishment of the Bell Telephone Company, which eventually became American Telephone and Telegraph (AT&T).

THE JOB

Because electrical engineering is such a diverse field, there are numerous divisions in which engineers work. In fact, the discipline reaches nearly every other field of applied science and technology, including renewable energy. In general, electrical engineers use their knowledge of the sciences in the practical applications of electrical energy. They specialize in power systems engineering and electrical equipment manufacturing. They are involved in the invention, design, construction, and operation of electrical systems and devices of all kinds.

The work of electrical engineers touches almost every niche of our lives. Think of the things around you that have been designed, manufactured, maintained, or in any other way affected by electrical energy: the lights in a room, cars on the road, televisions, stereo systems, telephones, your doctor's blood-pressure reader, computers. When you start to think in these terms, you will discover that the electrical engineer has in some way had a hand in science, industry, commerce, entertainment, and even art.

The list of specialties that engineers are associated with reads like an alphabet of scientific titles—from acoustics, speech, and signal processing; to electromagnetic compatibility; geoscience and remote sensing; lasers and electro-optics; robotics; ultrasonics, ferroelectrics, and frequency control; to vehicular technology. As evident in this selected list, engineers are apt to specialize in what interests them, such as communications, robotics, or automobiles. There are also electrical engineers who are environmental and emissions specialists. They design systems and run tests to make sure they comply with environmental and emissions limits set by the government.

As mentioned earlier, electrical engineers focus on high-power generation of electricity and how it is transmitted for use in lighting homes and powering factories. They are also concerned with how equipment is designed and maintained and how communications are transmitted via wire and airwaves. Some are involved in the design and construction of power plants and the manufacture and maintenance of industrial machinery.

Electrical engineers may design and maintain power "grids," as well as the power systems that connect to them. The United States, like most countries, maintains an electrical network that connects a variety of electric generators to electric-power users. Many users save money by purchasing power from this "grid" rather than having to generate the power themselves. Some electrical engineers may also work "off-grid," meaning on systems that are *not* connected to the grid. (Some users may find it cheaper to generate their own power.)

Paul Pabst is an electrical engineer who works in the electric power industry, which has several large divisions, including electricity generation, delivery, and protection. His specialty is in the protection of electrical systems. Paul designs "secure, functional, and unique power systems to assure that the delivery of power remains safe under any conditions." He explains that "these 'conditions' can range from lightning hitting a power line, a tree branch falling and destroying power lines, or any number of scenarios that cause hazards to the electrical system." He works with drawings that simulate the entire electrical system, and with protection fundamentals to create a protection scheme.

Other areas in which electrical engineers may find their niche include design and testing, research and development, production, field service, sales and marketing, and teaching. And even within each category there are divisions of labor.

Researchers concern themselves mainly with issues that pertain to potential applications. They conduct tests and perform studies to evaluate fundamental problems involving such things as new materials and chemical interactions. Those who work in design and development adapt the researchers' findings to actual practical applications. They devise functioning devices and draw up plans for their efficient production, using computer-aided design and engineering (CAD/CAE) tools. For a typical product such as a television, this phase usually takes up to 18 months to accomplish. For other products, particularly those that utilize developing technology, this phase can take as long as 10 years or more.

Production engineers have perhaps the most hands-on tasks in the field. They are responsible for the organization of the actual manufacture of whatever electric product is being made. They take care of materials and machinery, schedule technicians and assembly workers, and make sure that standards are met and products are quality controlled. These engineers must have access to the best tools for measurement, materials handling, and processing.

After electrical systems are put in place, *field service engineers* act as the liaison between the manufacturer or distributor and the client. They ensure the correct installation, operation, and maintenance of systems and products for both industry and individuals. In the sales and marketing divisions, engineers stay abreast of customer needs in order to evaluate potential applications, and they advise their companies of orders and effective marketing. A *sales engineer* would contact a client interested in, say, a certain type of microchip for its automobile electrical system controls. He or she would learn about the client's needs and report back to the various engineering teams at his or her company. During the manufacture and distribution of the product, the sales engineer would continue to communicate information between company and client until all objectives were met.

All engineers must be taught their skills, so it is important that some remain involved in academia. *Professors* usually teach a portion of the basic engineering courses as well as classes in the subjects in which they specialize. Conducting personal research is generally an ongoing task for professors in addition to the supervision of student work and student research. Part of the teacher's time is also devoted to providing career and academic guidance to students.

Whatever type of project an engineer works on, he or she is also likely to have a certain amount of desk work. Writing status reports and communicating with clients and others who are working on the same project are examples of the paperwork that most engineers are responsible for.

REQUIREMENTS
High School
Electrical engineers must have a solid educational background, and the discipline requires a clear understanding of practical applications. To prepare for college, high school students should take classes in algebra, trigonometry, calculus, biology, physics,

(continues on page 28)

INTERVIEW

Paul Pabst, Electrical Engineer

Q. How long have you been an electrical engineer, and what sparked your interest in this type of engineering?

A. I have been an electrical engineer for two years. I participated in an intense internship program that gave me roughly two years of electrical engineering experience before graduation. I have always been fascinated with using my creative mind in ways that I can't in normal, everyday ways. Engineering is something that requires a significant amount of technical knowledge, but it also requires creativity and innovation. Much of the work I do is unique with respect to other projects, and this requires me to quickly adapt to new ideas and develop unique solutions to the unique challenges.

I wasn't always sure that I wanted to be an electrical engineer. It took me several different majors, after deciding that I wanted to pursue engineering, to decide on electrical engineering. I really feel that gaining real-world work experience prior to graduation is what helped me solidify my decision to become an engineer. It was very satisfying to learn about the field of engineering by actually getting into it through the internship.

Q. What is your work and educational background?

A. I interned at this company and received a full-time position following my graduation. As an intern, I was in four different divisions of my company and was able to get a very wide spectrum of the different roles that engineers fulfill in the workplace. In my full-time role, I have spent the entire time in the same division, in my power systems role.

I went to a top-tier engineering school and received my BS in electrical engineering. Because of the internship (called a cooperative education program), my graduation was postponed one year, and I thus graduated in five years with roughly two years of work experience (summers were spent either working or in school).

Q. What do you like most about your work? And what do you like least?

A. I built a strong foundation of knowledge at school; however, the most useful tool I gained at school was "learning how to learn." I developed a very strong sense of how to learn new concepts, how to rely on my engineering fundamentals and welcome new ideas without fear. In my work, I encounter new ideas and concepts nearly every day and am generally very comfortable and confident that I will be able to handle whatever is thrown at me. This is what I enjoy most about my work—the opportunity to deal with these new ideas and new concepts. I am, in turn, constantly learning, which keeps me excited about what I do.

One of the biggest challenges that I face in my work is "underexperience." I am constantly being pushed to my technical limits and at times am lacking the abilities and experience to answer questions. I find that sometimes I am relied upon but cannot deliver. It is a concern of mine, but I am also confident that as long as I continue to be open-minded and develop myself as an electrical engineer, that I will continue to develop and hone my skills.

Q. What surprised you about this work?
A. The amount of teamwork is something I did not expect. There are misconceptions that engineers work mainly alone and present their findings to management at the end of the day. I found out very quickly that this is not the case. I am constantly working in teams, conversing with others, and coming to group conclusions on the best way to move forward. At times there are disagreements and debates on the direction of the project, but things are quickly resolved once the technical details are discussed and examined.

Q. What advice can you give to students who are considering electrical engineering as a career?
A. I cannot stress enough the importance of going to college and learning in a group environment while being mentored by some of the brightest people in the world. The second-best piece of advice is to try out electrical engineering via an internship before you graduate. The skills you learn in school will never exactly align with what you are going to do after you graduate, which is why it's very important to also gain real-world skills.

(continued from page 25)

chemistry, computer science, word processing, English, and social studies. Business classes are also helpful. And students who are planning to pursue studies beyond a bachelor of science degree will also need to take a foreign language. It is recommended that students aim for honors-level courses.

Postsecondary Training

Most electrical engineers have a bachelor of science in electrical engineering. Other engineers might receive similar degrees in electronics, computer engineering, or another related science. Electrical engineering programs vary from one school to another, so be sure to explore as many schools as possible to determine which program is most suited to your academic and personal interests and needs. Most engineering programs have strict admission requirements and require students to have excellent academic records and top scores on national college-entrance examinations. Competition can be fierce for some programs, and high school students are encouraged to apply early.

Many students go on to receive a master of science degree in a specialization of their choice. This usually takes an additional two years of study beyond a bachelor's program. Some students pursue a master's degree immediately upon completion of a bachelor's degree. Other students, however, gain work experience first and then take graduate-level courses on a part-time basis while they are employed. A Ph.D. is also available. It generally requires four years of study and research beyond the bachelor's degree and is usually completed by people interested in research or teaching.

By the time you reach college, it is wise to start thinking about the type of engineering specialty you might be interested in. In addition to the core engineering curriculum (advanced mathematics, physical science, engineering science, mechanical drawing, computer applications), students will begin to choose from the following types of courses: circuits and electronics; signals and systems; digital electronics and computer architecture; electromagnetic waves, systems, and machinery; communications; and statistical mechanics.

Certification or Licensing

All 50 states and the District of Columbia require engineers who offer their services to the public to be licensed as professional engineers (PEs). To be designated as a PE, engineers must have a degree from an engineering program accredited by the Accreditation Board for Engineering and Technology, four years of relevant work experience, and successfully complete the state examination.

Other Requirements

To be a successful electrical engineer, you should have strong problem-solving abilities, mathematical and scientific aptitudes, and the willingness to learn throughout your career. Engineers are innately curious. "When I was younger, I really did enjoy learning 'how' stuff worked," Paul Pabst says. "It is always interesting to me to understand what makes your flashlight stay on, or how your cellular phone can connect you to anyone in the world at any time, or how the computer can do what it does today."

Being open-minded and able to listen to others' opinions is also important. Pabst says, "There are always multiple solutions to any given problem, and one's solution may not always be the best solution." Most engineers work on teams with other professionals, so the ability to get along with others is essential. In addition, strong communication skills are critical because engineers need to be able to write clear reports and give oral presentations.

EXPLORING

People who are interested in the excitement of electricity can tackle experiments such as building a radio or central processing unit of a computer. Special assignments can also be researched and supervised by teachers. Joining a science club, such as the Junior Engineering Technical Society (JETS), can provide hands-on activities and opportunities to explore scientific topics in-depth. Student members can join competitions and design structures that exhibit scientific know-how. Reading trade publications, such as the *Pre-Engineering Times*, are other ways to learn about the engineering field. This magazine includes articles on engineering-related careers and club activities. *R&D Magazine* is also great for learning about industry news and award-winning inventions. Paul Pabst also recommends *Fast Company* (http://www.fastcompany.com).

Students can also learn more about electrical engineering by attending a summer camp or academic program that focuses on scientific projects as well as recreational activities. For example, the Delphian School in Oregon holds summer sessions for high school students. Students are involved in leadership activities and special interests such as mathematics, electricity, and energy. Sports and wilderness activities are also offered. The Michigan Technological

University Summer Youth Program focuses on career exploration in engineering, computers, electronics, and robotics. Geared to high school students, this academic program also offers arts guidance, wilderness events, and other recreational activities. (For further information on clubs and programs, visit the Web sites listed at the end of this article.)

EMPLOYERS

Approximately 153,000 electrical engineers are employed in the United States. Most work in architectural, engineering, and related services. Some work for business consulting firms, and manufacturing companies that produce electrical and electronic equipment, business machines, computers and data processing components, and telecommunications parts. Others work for consulting firms; public utilities; government agencies; and companies that make automotive electronics, scientific equipment, and aircraft parts. Some work as private consultants.

STARTING OUT

Many students begin to research companies that they are interested in working for during their last year of college or earlier. You can research companies by looking at company directories and annual reports, which are available at public libraries.

Employment opportunities can be found through a variety of sources. Many companies recruit engineers while they are still in college. Other companies have internship, work-study, or cooperative education programs from which they hire students who are still in college. Students who have participated in these programs often receive permanent job offers through these companies, or they may obtain useful contacts that can lead to a job interview or offer. Some companies use employment agencies and state employment offices, or they post job openings directly on their Web sites (usually in the Careers or Work for Us section). Companies may also advertise positions through advertisements in newspapers and trade publications. In addition, many newsletters and associations post job listings on the Internet.

ADVANCEMENT

Engineering careers usually offer many avenues for advancement. An engineer straight out of college will usually take a job as an entry-level engineer and advance to higher positions after acquiring some job experience and technical skills. Engineers with strong technical skills who show leadership ability and good communication skills may move into positions that involve supervising teams of engineers and making sure they are working efficiently. Engineers can advance from these positions to that of a *chief engineer*. The chief engineer usually oversees all projects and has authority over project managers and managing engineers.

Many companies provide structured programs to train new employees and prepare them for advancement. These programs usually rely heavily on formal training opportunities such as in-house development programs and seminars. Some companies also provide special programs through colleges, universities, and outside agencies. Engineers usually advance from junior-level engineering jobs to more senior-level positions through a series of promotions. Engineers may also specialize in a specific area once they have acquired the necessary experience and skills.

Some engineers move into sales and managerial positions, with some engineers leaving the industry to seek top-level management positions with other types of firms. Other engineers set up their own firms in design or consulting. Engineers can also move into the academic field and become teachers at high schools or universities.

EARNINGS

Starting salaries for all engineers are generally much higher than for workers in any other field. Entry-level electrical/electronics and communication engineers with a bachelor's degree earned an average of $55,292, according to a 2007 salary survey by the National Association of Colleges and Employers. Those with a master's degree averaged around $66,309 in their first jobs after graduation, and those with a Ph.D. received average starting offers of $75,982. The U.S. Department of Labor reports that the median annual salary for electrical engineers was $82,160 in 2008. The lowest-paid engineers earned less than $52,990 and the highest-paid engineers earned more than $125,810 annually.

Most companies offer attractive benefits packages, although the actual benefits vary from company to company. Benefits can include any of the following: paid holidays, vacations, and personal days, sick leave; medical, health, and life insurance; short- and long-term disability insurance; profit sharing; 401(k) plans;

retirement and pension plans; educational assistance; leave time for educational purposes; and credit unions. Some companies also offer computer purchase assistance plans and discounts on company products.

WORK ENVIRONMENT

In many parts of the country, a five-day, 40-hour workweek is still the norm, but it is becoming much less common. Many engineers regularly work 10 or 20 hours of overtime a week. Engineers in research and development, or those conducting experiments, often need to work at night or on weekends. Workers who supervise production activities may need to come in during the evenings or on weekends to handle special production requirements. In addition to the time spent on the job, many engineers also participate in professional associations and pursue additional training during their free time. Many high-tech companies allow flextime, which means that workers can arrange their own schedules within certain time frames.

Most electrical engineers work in fairly comfortable environments. Engineers involved in research and design may work in specially equipped laboratories. Engineers involved in development and manufacturing work in offices and may spend part of their time in production facilities. Depending on the type of work one does, there may be extensive travel. Engineers involved in field service and sales spend a significant amount of time traveling to see clients. Engineers working for large corporations may travel to other plants and manufacturing companies, both around the country and at foreign locations.

Engineering professors spend part of their time teaching in classrooms, part of it doing research in labs or libraries, and some of the time still connected with industry in some capacity.

OUTLOOK

Opportunities for electrical engineers are expected to grow slower than the average for all other occupations through 2016, according to the *Occupational Outlook Handbook*. Although demand for electronics should increase, competition from foreign countries may limit job growth for electrical engineers in the United States. The career outlook for electrical engineers with certain specialties and in certain industries is not entirely bleak, though. Firms that

>
> ## A Small Sampling of Modern-Day Inventions by Electrical Engineers
>
> - CD players
> - Cell phones
> - Computers
> - Fax machines
> - Fiber-optic communications
> - The Internet
> - Microprocessors
> - MRI (magnetic resonance imaging) machines

provide engineering and design services are expected to show the fastest growth. Also, electrical engineers specializing in renewable energy, such as wind and solar power technologies, may find more employment opportunities in the coming years, particularly as more governments invest money in alternative-energy research and development. Paul Pabst also points out that an aging workforce will give rise to more opportunities for younger engineers. "The job outlook is very promising in my industry specifically," he says, "as many older people in the industry will be retiring in the next 10 years."

Engineers will need to stay on top of changes within the electronics industry and will need additional training throughout their careers to learn new technologies. Economic trends and conditions within the global marketplace have become increasingly more important. In the past, most electronics production was done in the United States or by American-owned companies. This changed during the 1990s, and the electronics industry entered an era of global production. Worldwide economies and production trends will have a larger impact on U.S. production, and companies that cannot compete technologically may not succeed. Job security is no longer assured, and many engineers can expect to make significant changes in their careers at least once. Engineers who have a strong academic foundation, who have acquired technical knowledge and skills, and who stay up to date on changing technologies provide themselves with the versatility and flexibility necessary to succeed within the electronics industry.

FOR MORE INFORMATION

For information on the Summer at Delphi Youth Program for high school students, contact

    The Delphian School
    20950 SW Rock Creek Road
    Sheridan, OR 97378-9740
    Tel: 800-626-6610
    Email: summer@delphian.org
    http://www.summeratdelphi.org

For information on careers and educational programs, contact the following associations:

    Institute of Electrical and Electronics Engineers
    2001 L Street, NW, Suite 700
    Washington, DC 20036-4910
    Tel: 202-785-0017
    Email: ieeeusa@ieee.org
    http://www.ieee.org

    Electronic Industries Alliance
    2500 Wilson Boulevard
    Arlington, VA 22201-3834
    Tel: 703-907-7500
    http://www.eia.org

For information on careers, educational programs, and student clubs, contact

    Junior Engineering Technical Society
    1420 King Street, Suite 405
    Alexandria, VA 22314-2794
    Tel: 703-548-5387
    http://www.jets.org

For information on its program for high school students, contact

    Michigan Technological University Summer Youth Program
    Youth Programs Office
    1400 Townsend Drive
    Houghton, MI 49931-1295
    Tel: 906-487-1885
    http://www.mtu.edu

Energy Conservation Technicians

QUICK FACTS

School Subjects
Mathematics
Technical/shop

Personal Skills
Helping/teaching
Mechanical/manipulative

Work Environment
Indoors and outdoors
Primarily multiple locations

Minimum Education Level
Associate's degree

Salary Range
$26,580 to $41,100 to $68,460+

Certification or Licensing
Voluntary

Outlook
About as fast as the average

OVERVIEW

Energy conservation technicians identify and measure the amount of energy used to heat, cool, and operate a building or industrial process. They analyze the efficiency of energy use and determine the amount of energy lost through wasteful processes or lack of insulation. After analysis, they suggest energy conservation techniques and install any needed corrective measures.

HISTORY

At the start of the 20th century, energy costs were only a fraction of the total expense of operating homes, offices, and factories. Coal and petroleum were abundant and relatively inexpensive. Low energy prices contributed to the emergence of the United States as the leading industrialized nation as well as the world's largest energy consumer.

Because petroleum was inexpensive and could easily produce heat, steam, electricity, and fuel, it displaced coal for many purposes. As a result, the nation's coal mining industry declined, and the United States became dependent on foreign oil for half of its energy supply.

In 1973, when many foreign oil-producing nations stopped shipments of oil to the United States and other Western countries, fuel costs increased dramatically. In the 2000s political instability in the Middle East—where many of the top oil-producing countries are located—caused fuel prices to once again rise significantly. This uncertainty regarding supply and a growing awareness about environmental pollution have fueled the development of energy conservation techniques in the United States. The emphasis on discovering new sources of energy, developing more efficient methods and equipment to use energy, and reducing the amount of wasted energy has created a demand for energy conservation technicians.

THE JOB

Energy efficiency and conservation are major concerns in nearly all homes and workplaces. This means that work assignments for energy conservation technicians vary greatly. They may inspect homes, businesses, or industrial buildings to identify conditions that cause energy waste, recommend ways to reduce the waste, and help install corrective measures. When technicians complete an analysis of a problem in energy use and effectiveness, they can state the results in tangible dollar costs, losses, or savings. Their work provides a basis for important decisions on using and conserving energy.

Energy conservation technicians may be employed in power plants, research laboratories, construction firms, industrial facilities, government agencies, or companies that sell and service equipment. The jobs these technicians perform can be divided into four major areas of energy activity: research and development, production, use, and conservation.

In research and development, technicians usually work in laboratories testing mechanical, electrical, chemical, pneumatic, hydraulic, thermal, or optical scientific principles. Typical employers include institutions, private industry, government, and the military. Working under the direction of an engineer, physicist, chemist, or metallurgist, technicians use specialized equipment and materials to perform laboratory experiments. They help record data and analyze it using computers. They may also be responsible for periodic maintenance and repair of equipment. Some help in creating prototype versions of newly designed equipment, using computer-aided design and drafting (CADD) software and equipment.

In energy production, typical employers include solar energy equipment manufacturers, installers, and users; power plants; and process plants that use high-temperature heat, steam, or water. Technicians in this field work with engineers and managers to develop, install, operate, maintain, and repair systems and devices used for the conversion of fuels or other resources into useful energy. Such plants may produce hot water, steam, mechanical motion, or electrical power through systems such as furnaces, electrical power plants, and solar heating systems. These systems may be controlled manually by semi-automated control panels or by computers.

In the field of energy use, technicians might work to improve efficiency in industrial engineering and production line equipment. They also maintain equipment and buildings for hospitals, schools, and multifamily housing.

Technicians working in energy conservation typically work for manufacturing companies, consulting engineers, energy-audit firms, and energy-audit departments of public utility companies. Municipal governments, hotels, architects, private builders, and manufacturers of heating, ventilating, and air-conditioning equipment also hire them. Working in teams led by engineers, technicians determine building specifications, modify equipment and structures, audit energy use and the efficiency of machines and systems, and then recommend modifications or changes to save energy.

If working for a utility company, a technician might work as part of a demand-side management (DSM) program, which helps customers reduce the amount of their electric bill. Under DSM programs, energy conservation technicians visit customers' homes to interview them about household energy use, such as the type of heating system, the number of people home during the day, the furnace temperature setting, and prior heating costs.

Technicians then draw a sketch of the house, measure its perimeter, windows, and doors, and record dimensions on the sketch. They inspect attics, crawl spaces, and basements and note any loose-fitting windows, uninsulated pipes, and deficient insulation. They read hot-water tank labels to find the heat-loss rating and determine the need for a tank insulation blanket. Technicians also examine air furnace filters and heat exchangers to detect dirt or soot buildup that might affect furnace operations. Once technicians identify a problem, they must know how to correct it. After discussing problems with the customer, the technician recommends repairs and provides literature on conservation improvements and sources of loans.

REQUIREMENTS
High School
If you are interested in this field, take classes such as algebra, geometry, physics, chemistry, machine shop, and ecology. These courses and others incorporating laboratory work will provide you with a solid foundation for any postsecondary program that follows. In addition, classes in computer science, drafting (either mechanical or architectural), and public speaking are also very helpful.

Postsecondary Training
Most employers prefer to hire energy conservation technicians who have at least a two-year undergraduate degree. Many community colleges and technical institutes provide associate's degree programs under the specific title of energy conservation and use technology or energy management technology. In addition, schools offer related programs in solar power, electric power, building maintenance, equipment maintenance, and general engineering technology.

Advanced training focuses on the principles and applications of physics, energy conservation, energy economics, instrumentation, electronics, electromechanical systems, computers, heating systems, and air-conditioning. A typical curriculum offers a first year of study in physics, chemistry, mathematics, energy technology, energy production systems, electricity and electromechanical devices, and microcomputer operations. The second year of study becomes more specialized, including courses in mechanical and fluid systems, electrical power, blueprint reading, energy conservation, codes and regulations, technical communications, and energy audits.

Considerable time is spent in laboratories, where students gain hands-on experience assembling, disassembling, adjusting, and operating devices, mechanisms, and integrated systems of machines and controls.

Certification or Licensing
There are no state or federal requirements for certification or licensing of energy conservation technicians. However, certification from the National Institute for Certification in Engineering Technologies and a degree from an accredited technical program are valuable credentials and proof of knowledge and technical skills. The Association of Energy Engineers also offers various certifications to professionals in the field.

Other Requirements
Students entering this field must have a practical understanding of the physical sciences, a good aptitude for math, and the ability to communicate in writing and speech with technical specialists as well as average consumers. Their work requires a clear and precise understanding of operational and maintenance manuals, schematic drawings, blueprints, and computational formulas.

Some positions in electrical power plants require energy conservation technicians to pass certain psychological tests to predict their behavior during crises. Security clearances, arranged by the employer, are required for employment in nuclear power plants and other workplaces designated by the government as high-security facilities.

EXPLORING
To learn more about this profession, ask your career guidance counselor for additional information or for assistance in arranging a field trip to an industrial, commercial, or business workplace to explore energy efficiency.

Utility companies exist in almost every city and employ energy analysts or a team of auditors in their customer service departments. Energy specialists also work for large hospitals, office buildings, hotels, universities, and manufacturing plants. Contact these employers of energy technicians to learn about opportunities for volunteer, part-time, or summer work.

You can also enroll in seminars offered by community colleges or equipment and material suppliers to learn about such topics as building insulation and energy sources. Student projects in energy conservation or part-time work with social service agencies that help low-income citizens meet their energy costs are other options for exploration.

EMPLOYERS

Energy conservation technicians are employed in areas where much energy is used, such as power plants, research laboratories, construction firms, industrial facilities, government agencies, and companies that sell and service equipment. Technicians who focus on research and development work for institutions, private industry, government, and the military. Those who work in energy use are employed by manufacturing facilities, consulting engineering firms, energy audit firms, and energy audit departments of utility companies. Other employers include municipal governments, manufacturers of heating and cooling equipment, private builders, hotels, and architects.

STARTING OUT

Most graduates of technical programs are able to secure jobs in energy conservation before graduation by working with their schools' career services offices. Placement staffs work closely with potential employers, especially those that have hired graduates in recent years. Many large companies schedule regular recruiting visits to schools before graduation.

It is also possible to enter the field of energy conservation on the basis of work experience. People with a background in construction, plumbing, insulation, or heating may enter this field with the help of additional training to supplement past work experience. Training in military instrumentation and systems control and maintenance is also good preparation for the prospective energy conservation technician. Former navy technicians are particularly attractive job candidates in the field of energy production.

Opportunities for on-the-job training in energy conservation are available through summer or part-time work in hospitals, major office buildings, hotel chains, and universities. Some regions have youth corps aimed at high school students, such as the Northwest Youth Corps in Oregon, which offers education and on-the-job training opportunities to youth and young adults.

Some jobs in energy production, such as those in electrical power plants, can be obtained right out of high school. New employees, however, are expected to successfully complete company-sponsored training courses to keep up to date and advance to positions with more responsibility. Graduates with associate's degrees in energy conservation and use, instrumentation, electronics, or electromechanical technology will normally enter employment at a higher level, with more responsibility and higher pay, than those with less education. Jobs in energy research and development almost always require an associate's degree.

ADVANCEMENT

Because the career is relatively new, well-established patterns of advancement have not yet emerged. Nevertheless, technicians in any of the four areas of energy conservation can advance to higher positions, such as senior and supervisory positions. These advanced positions require a combination of formal education, work experience, and special seminars or classes usually sponsored or paid for by the employer.

Technicians can also advance by progressing to new, more challenging assignments. For example, hotels, restaurants, and retail stores hire experienced energy technicians to manage energy consumption. This often involves visits to each location to audit and examine its facilities or procedures to see where energy use can be reduced. The technician then provides training in energy-saving practices. Other experienced energy technicians may be employed as sales and customer service representatives for producers of power, energy, special control systems, or equipment designed to improve energy efficiency.

Technicians with experience and money to invest may start their own businesses, selling energy-saving products, providing audits, or recommending energy-efficient renovations.

EARNINGS

Earnings of energy conservation technicians vary significantly based on the amount of formal training and experience. According to the U.S. Department of Labor, the average annual salary for environmental engineering technicians in engineering and architectural services was $41,100 in 2008. Salaries for all environmental engineering technicians ranged from less than $26,580 to $68,460 or more annually.

Technicians typically receive paid vacations, group insurance benefits, and employee retirement plans. In addition, their employers often offer financial support for all or part of continuing education programs, which are necessary in order for technicians to keep up to date with technological changes that occur in this developing field.

WORK ENVIRONMENT

Because energy conservation technicians are employed in many different settings, the environment in which they work varies widely. Energy conservation technicians employed in research and development, design, or product planning generally work in laboratories or engineering departments with normal daytime work schedules. Other technicians often travel to customer locations or work in their employer's plant.

Work in energy production and use requires around-the-clock shifts. In these two areas, technicians work either indoors or outdoors at the employer's site. Such assignments require little or no travel, but the work environments may be dirty, noisy, or affected

No- or Low-Cost Ways to Save Energy

Here are some ideas that can help you save energy and money at your home or small business:

- Replace incandescent bulbs with compact fluorescent lights (CFLs).
- Air-dry dishes instead of using your dishwasher's drying cycle.
- Use your microwave instead of a conventional electric range or oven.
- Turn off your computer and monitor when not in use.
- Plug home electronics, such as TVs and DVD players, into power strips and turn power strips off when equipment is not in use.
- Lower the thermostat on your hot water heater; 115° is comfortable for most uses.
- Take showers instead of baths to reduce hot water use.
- Wash only full loads of dishes and clothes.
- Choose Energy Star appliances, which use energy more efficiently.

Source: U.S. Department of Energy. Visit http://www.energy.gov/energytips.htm for more tips on saving energy.

by the weather. Appropriate work clothing must be worn in shop and factory settings, and safety awareness and safe working habits must be practiced at all times.

Energy conservation technicians who work in a plant usually interact with only a small group of people, but those who work for utility companies may have to communicate with the public while providing technical services to their customers. Energy research and development jobs involve laboratory activities requiring social interaction with engineers, scientists, and other technicians. In some cases, technicians may be considered public relations representatives, which may call for special attention to dress and overall appearance.

Job stress varies depending on the job. The pace is relaxed but businesslike in engineering, planning, and design departments and in research and development. In more hectic areas, however, technicians must respond to crisis situations involving unexpected breakdowns of equipment that must be corrected as soon as possible.

OUTLOOK

Since energy use constitutes a major expense for industry, commerce, government, institutions, and private citizens, the demand for energy conservation technicians is likely to remain strong. The U.S. Department of Labor predicts that the employment of environmental engineering technicians is expected to increase as fast as the average for all occupations through 2016. In addition to the financial costs of purchasing natural resources, the added reality of the physical costs of depleting these important resources continues to create a greater demand for trained energy conservation employees. Keep in mind that employment is influenced by local and national economic conditions. During economic slowdowns, fewer jobs may exist and competition increases for those jobs.

The utilities industry is in the midst of significant regulatory and institutional changes. Government regulations are moving utility companies toward deregulation, opening new avenues for energy service companies. In the past, people with engineering and other technical skills have dominated energy conservation programs. These skills will remain important, but as the industry becomes more customer-focused, there will be a growing need for more people with marketing and financial skills.

Utility companies, manufacturers, and government agencies are working together to establish energy efficiency standards. The

Consortium for Energy Efficiency (http://www.cee1.org) is a collaborative effort involving a group of electric and gas utility companies, government energy agencies, research organizations, and environmental groups working to develop programs aimed at improving energy efficiency in commercial air-conditioning equipment, lighting, geothermal heat pumps, and other systems. Programs such as these will create job opportunities for technicians.

Utility DSM programs, which have traditionally concentrated on the residential sector, are now focusing more attention on industrial and commercial facilities. With the goal of realizing larger energy savings, lower costs, and more permanent energy-efficient changes, DSM programs are expanding to work with contractors, builders, retailers, distributors, and manufacturers, creating more demand for technicians.

FOR MORE INFORMATION

This trade association represents employees in the petroleum industry. For free videos, fact sheets, and informational booklets available to educators, contact

American Petroleum Institute
1220 L Street, NW
Washington, DC 20005-4070
Tel: 202-682-8000
http://www.api.org

For information on technical seminars, certification programs, conferences, books, and journals, contact

Association of Energy Engineers
4025 Pleasantdale Road, Suite 420
Atlanta, GA 30340-4260
Tel: 770-447-5083
Email: info@aeecenter.org
http://www.aeecenter.org

For information on certification programs for engineering technicians and technologists, contact

National Institute for Certification in Engineering Technologies
1420 King Street
Alexandria, VA 22314-2794
Tel: 888-476-4238
Email: tech@nicet.org
http://www.nicet.org

For information on energy efficiency and renewable energy, visit the following Web site:

U.S. Department of Energy
Energy Efficiency and Renewable Energy
http://www.eere.energy.gov

Geotechnical Engineers

 QUICK FACTS

School Subjects
Mathematics
Technical/shop

Personal Skills
Mechanical/manipulative
Scientific/technical

Work Environment
Indoors and outdoors
One or more locations

Minimum Education Level
Bachelor's degree

Salary Range
$53,641 to $89,070 to $107,888+

Certification or Licensing
Required for certain positions

Outlook
Faster than the average

OVERVIEW

Geotechnical engineers research and study soil to evaluate its suitability for foundations. They investigate and assess construction sites, conduct lab tests, create designs for structures, supervise construction, and write and present reports. They work on such projects as designing tunnels and roadways, retaining walls and earth dams, as well as help create strategies for the clean-up and management of contaminated sites.

HISTORY

Geotechnical engineering is a discipline of civil engineering, the oldest engineering discipline after military engineering. Though the title did not exist around 2700 B.C., the builders of the pyramids in Egypt could be considered early civil engineers. Other ancient civil-engineer achievements include the Parthenon, built in Greece in the 5th century; the 4,000-mile Great Wall of China, built and rebuilt multiple times from the 5th century B.C. through the 1600s; and the Romans' aqueducts, harbors, waterways, dams, and roadways, which were used as early as the 7th century B.C.

It wasn't until the 18th century that civil engineering was distinguished as a discipline separate from military engineering. John Smeaton, a British engineer and physicist, was the first self-proclaimed civil engineer. In the 1750s, he designed the third Eddystone Lighthouse, an especially challenging project because the lighthouse is set on treacherous rocks in Devon, England. Using what today might be considered geotechnical engineering, Smeaton explored types of materials that would set under water for the foundation of the lighthouse, and discovered that hydraulic lime worked best. This discovery was used as the basis for the later development of cement. Because of Smeaton's design and choice of materials, the Eddystone Lighthouse withstood the elements for 100 years.

Civil engineering was officially recognized as a profession in 1828, with a Royal Charter by the Institution of Civil Engineers (founded in 1818, in London). The formal definition was given as follows:

> "... the art of directing the great sources of power in nature for the use and convenience of man, as the means of production and of traffic in states, both for external and internal trade, as applied in the construction of roads, bridges, aqueducts, canals, river navigation and docks for internal intercourse and exchange, and in the construction of ports, harbours, moles, breakwaters and lighthouses, and in the art of navigation by artificial power for the purposes of commerce, and in the construction and application of machinery, and in the drainage of cities and towns."

In 1819 Norwich University in Northfield, Vermont, was the first private college in America to offer classes in civil engineering. In 1835 Rensselaer Polytechnic Institute became the first U.S. school to award a civil engineering degree. And in 1905 Nora Stanton Blatch, a Cornell University student, was the first woman to receive a degree in civil engineering.

Engineering has grown since the 20th century to encompass multiple disciplines and specialties, many of which are now needed to address national infrastructure issues and global environmental concerns. Today, civil engineering, geotechnical engineering, and other engineering degree programs are offered at schools throughout the country and around the world.

THE JOB

Geotechnical engineers evaluate soil, rock, and underground water to determine the feasibility and design specifications for construction projects. Geotechnical engineering, like civil engineering, focuses on the design and construction of buildings, roads, tunnels, dams, bridges, and water supply and sewage systems. Types of problems geotechnical engineers may be hired to solve include leaning towers and buildings, railway track failures, and slope failures and landslides. Because geology plays a heavy role in this type of work, geotechnical engineers might also be known as *soil engineers*, *ground engineers*, or *geotechnics engineers*.

Before construction can be done on a building or a roadway, for instance, geotechnical engineers need to test the soil to determine what materials are safe to be mixed with it to create the foundation. They also determine if the soil can even be used, and if the site is appropriate and safe for construction. Geotechnical engineers use special drills to collect soil samples, which they then test for ground strength. They analyze the amount of rock, sand, clay, and moisture that's present in the soil sample. They also perform tests for soil pressure and composition. They then write a report on the soil composition and their recommendations for construction and development at the site.

Challenges geotechnical engineers face can include building on swamps or hills, renovating sites that are prone to flooding, or even coming up with strategies to clean up sites that are hazardous. They often work closely with civil and structural engineers, as well as with developers, landscapers, contractors, construction crews, and sometimes landowners. They usually work in a geotechnical consulting firm, with a team composed of project directors or managers, senior engineers, geologists, checkers and/or reviewers (making sure the project meets standards and specifications), as well as other civil and geotechnical engineers.

Other geotechnical engineering job responsibilities can include planning site investigations, creating foundation and structure designs using software design programs, managing projects, creating project estimates and budgets, preparing construction documentation, and working closely with and managing clients. Some geotechnical engineers with more years of experience may work in the area of identifying sites that have potential for development and renovation. For most geotechnical engineers, some degree of travel—it may be local or international, depending on the project—is required to conduct analysis, meet with teams, and oversee work.

REQUIREMENTS
High School
While in high school, take as many math and science classes as possible, including algebra, geometry, trigonometry, calculus, physics,

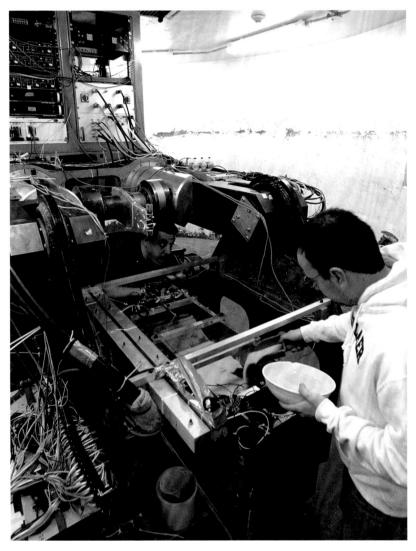

A geotechnical engineer (*right*) works with a professor of civil and environmental engineering at Rensselaer Polytechnic Institute to prepare a model of a New Orleans levee. The model, which sits in the box in the middle of the photo, will be spun to replicate the weather conditions during Hurricane Katrina. *AP Photo/Jim McKnight*

chemistry, and geology. Shop and drafting classes are also important. Engineers write a lot of reports and work with a variety of people, so strong communication skills are essential—communications and English classes will help you in this area.

Postsecondary Training

Geotechnical engineers may have a bachelor of science degree in geotechnical engineering, civil engineering, or structural engineering. They take such courses as hydrogeology, geology, construction engineering, environmental engineering, solid waste management, groundwater hydrology, pavement design, rock fragmentation, and rock-slope stability. Computer science classes are also strongly recommended, and business management classes are extremely useful. Most engineers gain practical work experience while in college by apprenticing with a geotechnical engineering firm.

Many companies prefer to hire engineers with a master of science degree in engineering.

Certification or Licensing

In the United States, a professional engineer (PE) license is required for engineers who provide services to the public. To receive a PE designation, engineers must have a degree from an ABET-accredited engineering program, four years of related work experience, and pass a state examination.

Engineers who secure voluntary certification can advance to positions of greater responsibility and secure higher salaries. Organizations such as the American Academy of Environmental Engineers (http://www.aaee.net) and the Institute of Electrical and Electronics Engineers (http://www.ieeeusa.org) offer certification programs to professional engineers.

Other Requirements

Geotechnical engineers enjoy investigating sites and solving problems. They write many reports, often translating technical information into language that people less familiar with the industry will understand. Strong analytical and creative abilities combined with excellent written and verbal communication skills are essential in this field. Knowledge of different computer programs, databases, and soil-analysis software is also required.

EXPLORING

Visit the Geotechnical Engineering section of the U.S. Department of Transportation's Web site (http://www.fhwa.dot.gov/

engineering/geotech/) to learn more about technologies, innovations, current projects, and to find publications and workshop listings. Read magazines and books about geotechnical engineering to gain a better understanding of the field. Another great way to learn firsthand about geotechnical engineering projects currently underway is to visit the Web sites of engineering consulting firms. Use Google and key in the words "geotechnical engineering projects" and see what comes up.

EMPLOYERS

The types of companies that hire geotechnical engineers include geotechnical and engineering consulting firms, public utilities, governmental agencies, and environmental organizations.

STARTING OUT

An internship or apprenticeship is the best way to learn more about this type of work. Search the Internet for geotechnical and civil engineering firms in your area, and see if they have any openings for summer or part-time help. College professors may also be looking for research help on projects. Look through the engineering sections of college sites to see if any listings are posted. You can also visit these Web sites to get the inside scoop on engineering projects and find listings for upcoming job fairs and job openings: GeoPrac (http://geoprac.net) and iCivilEngineer (http://iCivilEngineer.com).

ADVANCEMENT

Engineers usually start as associates and advance to become managers or directors. With a few years of successful project management experience, they can move up to regional management positions—overseeing more projects and staff at various work sites. They may also advance by honing skills in specific areas of engineering and becoming technical specialists. They can also start their own consulting firms, teach at universities, and mentor and coach less-experienced engineers and students.

EARNINGS

According to Salary.com, geotechnical engineers with some work experience averaged $58,871 per year in 2009—the bottom 10 percent earned $53,641 and the top 10 percent earned $63,120 or more. More experienced geotechnical engineers had higher annual salaries, ranging from $89,070 to $107,888. The U.S. Department of Labor reports that annual salaries for civil engineers in 2008 ranged from $48,140 to $115,630 or higher.

Engineers also receive employment benefits including 401(k), health care and disability insurance coverage, paid time off, social security, and pensions.

WORK ENVIRONMENT

Geotechnical engineers work in offices and outdoors at construction sites. They usually work 40 hours per week, but more hours are required when projects are nearing final deadlines. They also travel to different sites to oversee projects and consult with staff and clients.

OUTLOOK

Through 2016 civil engineers are expected to have faster than average employment growth, according to the U.S. Department of Labor (DoL). Population growth coupled with the country's aging infrastructure will spark the need for more civil engineers. Projects they will be needed for include designing and constructing or expanding transportation, water supply, and pollution-control systems, as well as buildings and building complexes. Civil and geotechnical engineers will also be needed to renovate or replace roads, bridges, or other structures.

Tighter environmental regulations and the country's increased focus on health concerns are spurring greater demand for environmental engineers. The DoL predicts that environmental engineers will experience much faster than average employment growth through the next decade.

FOR MORE INFORMATION

Find career resources for civil engineering at the ASCE's Web site.
  American Society of Civil Engineers (ASCE)
  1801 Alexander Bell Drive
  Reston, VA 20191-5467
  Tel: 800-548-2723
  http://www.asce.org

For industry news and events, legislative updates, and job links, visit
  California Geoprofessionals Association
  PO Box 1693

Placerville, CA 95667-1693
Tel: 530-344-0644
http://www.cgea.org

Find scores of publications on civil and geotechnical engineering at the following Web site:
Institution of Civil Engineers
http://www.icevirtuallibrary.com/content/journals

For a global perspective on the geotechnical engineering industry, visit
International Society for Soil Mechanics and Geotechnical Engineering
City University, Northampton Square
London EC1V 0HB
United Kingdom
Email: secretariat@issmge.org
http://www.issmge.org

Green Vehicle Designers

QUICK FACTS

School Subjects
Math/science
Technical/shop

Personal Skills
Mechanical/manipulative
Technical/scientific

Work Environment
Primarily indoors
One location

Minimum Education Level
Bachelor's degree

Salary Range
$31,400 to $57,350 to $97,770+

Certification or Licensing
Not required

Outlook
About as fast as the average

OVERVIEW

Green vehicle designers help create vehicles that have less impact on the environment than conventional cars. Types of vehicles they design include those that have fewer carbon emissions, use alternative fuel, or are powered fully or partially by electricity. In addition to designing the vehicles, they may also design systems that can be used to convert existing vehicles into "green" machines. They work independently as well as with other designers and engineers, manufacturers, business owners, and customers.

HISTORY

In the early days of motoring, electricity and steam were favored power sources. The electric vehicle actually has its roots in the electromagnetic cart that Dutch Inventor Sibrandus Stratingh created in 1835. He based the design on a steam engine he'd created the previous year. Inventors and scientists experimented with Stratingh's cart over the years until, by the late 1800s, passenger vehicles had evolved that were powered by rechargeable batteries and capable of moving at low speeds. By this time, also, steam locomotives had inspired twins Francis Stanley and Freelan Stanley to design and manufacture steam engine vehicles. The cars were marketed as the Stanley Steamers and were popular from 1896 until the late 1910s, when more fuel-efficient and powerful internal combustion engines replaced them.

Petroleum and diesel engines ruled the first part of the 20th century. In 1907 Henry Ford introduced the Model T, which cost half the price of an electric vehicle and could travel farther and faster. German Inventor Rudolf Diesel created the first diesel engine—which, believe it or not, was originally designed to run on vegetable oil—in 1893. Citroen and Mercedes-Benz were the first companies to produce commercial cars with diesel engines, and the debate continues (sometimes heatedly) regarding who, between the two, was the *very* first.

People again started tinkering around with green vehicles in the 1950s and 1960s. Between 1957 and 1961 the motorbike company Vespa produced the Vespa 400, a very successful two-seat, two-cylinder-engine microcar. Other small-engine cars of that era include the Austin-Healey Sprite and the Mini, which was produced until 2000 and has enjoyed a film career as well, being featured in such movies as *The Italian Job*, *The Bourne Identity*, and *Lara Croft: Tomb Raider*.

The oil crisis of the 1970s renewed interest in vehicles that used alternative fuel. Since then, researchers, inventors, and designers around the world have been developing vehicles that are fueled by electricity, solar energy, and biofuels (e.g., vegetable oils and animal fats). Today's hybrid vehicles are a result of all of this research and design. The Hybrid Center, a division of the Union of Concerned Scientists, defines a hybrid vehicle as follows: "A hybrid electric vehicle combines an internal combustion engine and an electric motor powered by batteries, merging the best features of today's combustion engine cars and electric vehicles. The combination allows the electric motor and batteries to help the conventional engine operate more efficiently, cutting down on fuel use. Meanwhile, the gasoline-fueled combustion engine overcomes the limited driving range of an electric vehicle. In the end, this hybridization gives you the ability to drive 500 miles or more using less fuel and never having to plug in for recharging."

THE JOB

New emissions standards are driving automakers, vehicle designers and engineers, environmentalists, and federal agencies to work

INTERVIEW

Pete Hansen, Green Vehicle Designer

Pete Hansen, president and CEO of Evolve-it Motors (http://www.evolveitmotors.com), has been in the green car business for three years. He is also head engineer and mechanic. The company specializes in converting vehicles to hybrid-electric or full-electric vehicles.

Q. What type of green vehicle design work do you do?
A. We work on systems, integration, fabrication, some vehicle-modification work, and installation. We collaborate with companies such as Braille Battery, Universal Battery, R.M.S, Eetrex, and Beck. Our customers have ranged from individuals who want their personal or family vehicles converted, to businesses with fleets (meaning a number of vehicles that are used for that particular business).

Q. What are the main skills needed to do this kind of work?
A. Knowledge of physics is helpful, as is having any kind of machine operation knowledge (including CAD), and extensive knowledge of electronics and mechanics. You must also have extremely good computer skills.

Q. What are you working on now that's new and exciting?
A. We're working with Beck, and we are now selling new electric vehicles, including the Beck Spyder and Beck 356 Speedster +e. We very much enjoy working on this project because it's the

together to create vehicles that have less impact on the environment. Green vehicles come in many different shapes, sizes, and colors, and they use a variety of alternative energy sources for fuel. They may be powered by compressed natural gas, biofuels, biodiesel, ethanol, clean diesel, fuel cell-powered hydrogen, hybrid gasoline electric, or full (plug-in) electric. They may be cars, minivans, SUVs, trucks, or motorcycles, used for commercial

Green Vehicle Designers   57

first time we've been able to offer a new car design with our system.

Q. What do you enjoy most about your job? And what's your least favorite part?
A. I like the challenge of getting any gas-powered vehicle and converting it to clean electric power, while keeping its integrity. What I don't like is how difficult it is to get parts needed quickly, and at a reasonable price.

Q. When did you become a green vehicle designer, and why?
A. I became a green vehicle designer in 2006, after gas prices rose so dramatically. I got tired of having only one choice, and not wanting just that one choice, I decided to make a change. I began to research what could be done to make a big change. I began to research batteries and what runs only on clean batteries. I started with an electric skateboard that got up to 35 miles per hour. Then I started with an electric bicycle design. Finally, I began to move forward to cars and trucks.

Q. What's your work background?
A.
- Served in the U.S. Air Force as a SATCOMM Technician
- Worked as electrician/lighting designer/operations manager for concerts with touring bands and set up for stage theaters
- Machine operator for a water drilling company
- Machine fabricator and operator for a bullet-case manufacturing company

and business purposes, or for everyday commuting, errands, and family vacations.

Green vehicle designers and engineers work together to optimize design specifications. They may create vehicles from scratch, or create systems that will convert vehicles to be more green. When creating vehicles and systems, they take into account weight, range, fuel, cargo capacity, safety, aerodynamics, and ergonomics.

Manufacturers, car companies, and customers provide feedback about what they want most in and from the vehicle and what features they can do without. To gather this information, designers may meet with clients, conduct market research, read design and consumer publications, attend trade shows and conferences, and meet with prospective manufacturers, vendors, and users.

Once they have all the input they need, designers start putting ideas together in sketches they create by hand and via computer-aided design (CAD) and computer-aided industrial design (CAID) software, collaborating throughout the process with creative directors and product development teams. They might also create models out of clay, wood, metal, and other materials. They present their sketches, graphics, and models to clients, get their feedback, and then incorporate the suggested changes into the design.

Product safety is an important aspect of the work. Designers work closely with engineers, accountants, and cost estimators to determine if it's possible to make products safer while still being economical to manufacture. Numerous usability and safety tests are done to ensure safety, correct flaws, and meet specifications and standards.

REQUIREMENTS
High School
Take classes in math, science, physics, history, English, art, graphic design, and computer software programs. Understanding how machines work is an important part of the job, so take any and all electronics, mechanics, and shop classes that your school offers.

Postsecondary Training
Most commercial and industrial design positions require a bachelor's degree. Green vehicle designers usually have a bachelor's degree in industrial design, transportation design, or engineering. Course work usually includes principles of design, sketching, computer-aided design, industrial materials and processes, manufacturing methods, as well as classes in engineering, physical science, math, psychology, and anthropology. Many school design programs also include internships at manufacturing or design firms.

Green vehicle designers who have a master's degree in industrial design may have better chances of landing jobs. More schools are offering programs in environmental design and engineering. Designers with professional work experience also seek degrees in other

Tesla Motors CEO, chairman, and product architect Elon Musk (*left*) and chief designer Franz von Holhauzen pose for photos at the unveiling of the Tesla Model S all-electric five-door sedan. *AP Photo/Reed Saxon*

areas, such as business management, marketing, or engineering, to enhance their skills in the workplace.

Other Requirements

People who work as designers draw from both sides of the brain: right side for creative, left side for technical. They have a good visual eye, the ability to hand-sketch as well as use CAD to create designs, and they have solid knowledge of how the products being designed actually function. Problem solving and thinking "outside of the box" are also intrinsic to the work, according to Pete Hansen. He says that green vehicle designers "ask a lot of questions," and they have to "be tenacious in finding the answers."

EXPLORING

The media is flooded with information about green vehicles and alternative fuels. On any given day, you'd be hard pressed to pick up a newspaper and not see at least one article about a new development in green car design and production. Pay attention to it all. Read and absorb everything you can to keep up with the latest breakthroughs and designs. You can conduct your own search for the greenest vehicles through the Environmental Protection Agency's "Green Vehicle Guide," found at http://www.epa.gov/greenvehicles/Index.do.

Magazines such as *Newsweek* and *Time* can give you the business perspective of the green vehicle industry, while *Popular Mechanics, Green Car Magazine,* and *Car and Driver* focus on the design and engineering as well as the business side. Documentaries are another excellent source of information. To learn more about the history and plight of the electric car, check out *Who Killed the Electric Car?* (narrated by Martin Sheen, 2006). And if you haven't already seen it, watch *An Inconvenient Truth* (Al Gore, 2006) to learn more about the impact of greenhouse gas emissions on the environment.

EMPLOYERS

According to the U.S. Department of Labor, about 48,000 commercial and industrial designers were employed in the United States in

What Exactly Is a "Green Car"?

The Swedish Road Administration defines a green car as follows:

- Conventional cars: Petrol and diesel cars with carbon dioxide emissions that do not exceed 120 grams/kilometer.
- Alternative fuel cars: Cars that can run on fuels other than petrol or diesel and with fuel consumption that does not exceed 0.92 liter petrol/10 kilometers, 0.84 liter diesel/10 kilometers, or 0.97 cubic meter gas/10 kilometers.
- Electric cars: A passenger car meeting environmental class Mk EL standards and with electric energy consumption that does not exceed 3.7 kilowatt hours/10 kilometers.

Source: Government Offices of Sweden (http://www.sweden.gov.se/sb/d/8202/a/79866)

2006. Of those, about 30 percent were self-employed, and approximately 15 percent worked for engineering or specialized design services firms. The rest worked for manufacturing firms and service-providing companies.

Business owners and fleet owners also need and hire green vehicle designers, according to Pete Hansen. A fleet is composed of five or more vehicles that are registered under the same name and assigned an identifier code by the Department of Licensing of that particular state. "For instance, if a company has 30 vehicles running for them," Hansen says, "then all 30 can be converted to electric running vehicles."

STARTING OUT

Many green vehicle designers get their start by interning with companies that are researching, designing, and/or manufacturing green vehicles. This usually happens while they are college students. See if you can find an internship or part-time or summer job with a company near you. And in the meantime, you can start designing and experimenting with motors and machines on your own to see if this work suits you. "The harnessing of electrical energy is one great achievement!" Pete Hansen says. "You can build your own electric car out of a gas-guzzler. Get yourself an electric car conversion manual. You can purchase *Electricity4Gas* (http://www.electricity-4gas.com), which will teach you everything you need to know about powering your own car with electricity. It's a step-by-step manual that guides you in creating your electric car in your own garage or backyard. Every day people are converting their cars to electric for clean fuel."

ADVANCEMENT

Green vehicle designers can advance in many ways in their careers. Junior designers with several years of experience can move up to become associate designers or designers. From there, the path can lead to senior designer, chief designer, design department head, or other senior management positions. Some green vehicle designers expand into other design and engineering specialties by seeking advanced degrees and by taking on new clients and projects. Writing for magazines, books, and the Internet is another way to branch out in the field. The media is often looking for expert sources, and some green vehicle designers become known, reliable consultants for news stories. Those who work on staff with companies can move on by starting their own design and consulting firms. They may also teach in design schools or in colleges and universities.

EARNINGS

The U.S. Department of Labor reports that commercial and industrial designers had median annual incomes of $57,350 in 2008, with the lowest paid 10 percent averaging $31,400 and the top paid 10 percent bringing home $97,770 or more per year. Salaries may be higher for those who own and run their own company or who manage larger manufacturing companies or corporations. States with the highest concentration of commercial and industrial designers in 2008 were Michigan, Rhode Island, Vermont, Wisconsin, and Minnesota. In that same year, the top-paying states for designers were California, Minnesota, Michigan, Massachusetts, and Utah.

WORK ENVIRONMENT

Green vehicle designers usually work indoors in the offices of manufacturing companies, corporations, or design firms. Work hours are generally 40 hours per week, but designers who own their own companies or have positions of greater responsibility may put in more time. For example, to accomplish his duties as CEO of Evolve-It Motors, Pete Hansen often works 70 hours or more per week, and he is often not alone. "Everyone that is a part of the team puts in an effort and takes great part in our goals for our future," he says. "We have a peaceful environment, and everyone who works with us believes in the positive changes we are making for our future and our children's future."

Designers who are working on special projects, contracted for specific work, or under particularly tight deadlines will extend their hours to weeknights and weekends to accomplish the requirements of the job. They also travel to testing facilities, design centers, clients' exhibit sites, manufacturing facilities, and to other locations.

OUTLOOK

Employment growth for commercial and industrial designers is expected to be about as fast as the average for all occupations through 2016. This is an exciting time for green vehicle research and development, though. Green vehicle designers will be needed to meet the population's growing demand for fuel-efficient, environmentally friendlier vehicles, and to meet the government's requirement to lower carbon emissions. Competition will be keen, however. Green vehicle designers who are extremely skilled in computer-aided design, have strong academic backgrounds, and relevant work experience will have the advantage in the job market.

FOR MORE INFORMATION

Find information about energy policies, conferences, and workshops at
American Council for an Energy-Efficient Economy
529 14th Street, NW, Suite 600
Washington, DC 20045-1000
Tel: 202-507-4000
Email: info@aceee.org
http://www.aceee.org

Find industry trends, green car ratings, and more by visiting
Greener Cars
http://greenercars.org

Learn more about hybrid vehicles and industry trends by visiting
The Hybrid Center
c/o Union of Concerned Scientists
Two Brattle Square
Cambridge, MA 02238-9105
Tel: 617-547-5552
http://www.hybridcenter.org

For membership information and listings of upcoming events, visit
Industrial Designers Society of America
45195 Business Court, Suite 250
Dulles, VA 20166-6716
Tel: 703-707-6000
http://www.idsa.org

Follow the progress of an actual green vehicle design project at the Open Source Green Vehicle Project section of this Web site:
Society for Sustainable Mobility
http://www.osgv.org/open-design-model

Read up on energy-efficiency initiatives and find more information on alternative fuel sources at the Transportation section of the DoE's Web site.

U.S. Department of Energy (DoE)
http://www.energy.gov/energyefficiency/transportation.htm

Hydroelectric Engineers

QUICK FACTS

School Subjects
Mathematics
Technical/shop

Personal Skills
Helping/teaching
Mechanical/manipulative

Work Environment
Indoors and outdoors
One or more locations

Minimum Education Level
Bachelor's degree

Salary Range
$44,484 to $82,160 to $125,810+

Certification or Licensing
Required for certain positions

Outlook
About as fast as the average

OVERVIEW

Hydroelectric engineers are responsible for developing and maintaining the mechanical technology at hydroelectric plants and overseeing the overall performance of the plant. They create designs and specifications for hydroelectric projects, which can include details for structures and/or machinery and equipment. They work with or manage other engineers and technicians, and may travel to facilities to oversee operations.

HISTORY

Hydropower is one of the oldest renewable energy sources known to man. More than 2,000 years ago, the Greeks used waterpower to turn waterwheels for milling wheat into flour. Early settlers in the United States took great advantage of the abundant, flowing rivers across the states to fuel grindstones and mills. By the 1700s Americans were using hydropower to mill lumber and grain, and to pump irrigation water. Early waterwheels had flat panels, which worked best on larger rivers. Cups were later added to the panels to improve the waterwheel's efficiency in handling fast-moving and less-voluminous waterways.

Hydroelectric power is the energy produced from the gravitational force of falling or flowing water. The water is usually dammed up to create greater pressure, and released to drive generators and hydraulic turbines. The amount of energy harnessed depends on the volume of water and the height (known as the "head") from which it falls. The oldest hydropower site in the United States is believed to be in Stottville, New York, at Claverack Creek; the earliest waterwheel to generate electricity was installed there in 1871. The first commercial hydroelectric power—a waterwheel in Appleton, Wisconsin's, Fox River—came about in 1882 and supplied light to one house and two paper mills.

Niagara Falls is not only a great natural wonder, but for many years it was also the largest hydroelectric power station in the world. It began operating locally in 1891: An 86-foot cascade of water generated electricity to power nearby mill machinery and village streetlights. By 1896 the water from the Niagara River was transmitting power to Buffalo, New York, about 26 miles away. Today, the Niagara Power Project is located four and a half miles downstream from the Falls. It features two main facilities—the Robert Moses Niagara Power Plant and the Lewiston Pump-Generating Plant. A "forebay" (a reservoir or canal), which sits between the two plants, can hold about 740 million gallons of water; and a 1,900-acre reservoir situated behind the Lewiston Plant holds additional liquid fuel supplies. Up to 375,000 gallons a second are diverted from Niagara River and sent through conduits under the City of Niagara Falls to Lewiston. From there, water flowing through the Robert Moses Plant spins turbines that power generators, which then convert the mechanical energy into electrical energy.

By the 1900s roughly 40 percent of all electricity in the United States was powered by hydroelectric energy. With the growth in demand for hydroelectric power, the government started to get involved in regulating the development and management of hydroelectric power resources. The first Federal Water Power Act came about in 1901 to limit the rights to purchase public lands for hydropower sites. One year later, the United States Bureau of Reclamation (USBR) (formerly known as the United States Reclamation Service) was founded to oversee water resource management, specifically relating to the oversight and/or operation of numerous water diversion, delivery, and storage and hydroelectric power generation projects throughout the western United States. From 1905 through 1911 the USBR built the Roosevelt Dam in Arizona, the largest, and last, masonry dam in the United States (and it's still in operation today). The Water Power Act of 1920 created the Federal Power Commission, which would control hydroelectric dam construction on all navigable rivers. To further develop large-scale water projects, the Tennessee Valley Authority was created in the southeastern United States in 1933, and four years later the Bonneville Power Administration was established in the Pacific Northwest. Some of these large-scale projects include the Grand Coulee Dam in Washington State, the Shasta Dam in California, and the Marshall Ford Dam in Texas.

Hydroelectric power usage began to fade toward the end of the 20th century as nuclear and fossil-fuel plants grew. Controversy also arose regarding some of the downfalls of hydroelectric power plants, including their impact on land and wildlife habitat, displacement of local populations, and destruction of historically and culturally significant sites. Today, while some countries produce much of their power from hydroelectricity (e.g., nearly 99 percent of Norway's electricity derives from hydroelectric power), the United States is producing only 10 percent of its power from hydroelectricity. There is a resurgence of interest in hydroelectric power as an alternative energy source, however, and more research is being conducted into ways to design and implement it so that it will benefit people and do less harm to the environment.

THE JOB

Hydropower represents 19 percent of the total electricity production in the world, according to the U.S. Geological Survey. And there are still plenty of hydro resources yet to be developed—about two-thirds of the world's potential remains untapped. While long lead times are required to research and propose new hydroelectric power sites, and the costs to create facilities can be prohibitive, there are many aspects of hydroelectric power that make it an attractive alternative to fossil fuels. No fuel is burned, thus causing minimal pollution and reducing greenhouse gas emissions. Once built, costs to operate and maintain hydroelectric facilities are relatively low. And nature provides the water: rainfall renews the supply.

Hydroelectric projects supply power to public electricity networks as well as to industrial enterprises. Some hydroelectric facilities are built to provide electricity to aluminum electrolytic plants. For example, the Grand Coulee Dam, constructed between 1935 through 1941 in the Columbia River Basin, contributed to the World War II effort by providing power to Alcoa Aluminum in Bellingam, Washington, for airplane manufacturing. After the war, it continued providing power to aluminum industries, as well as irrigation and power to communities.

A site manager points to a scale model for a proposed project to build hydroelectric dams. Hydroelectric power has emerged as an attractive alternative to fossil fuels in recent years. *AP Photo/Joel Page*

Hydroelectric engineers who work for hydroelectric generation facilities review customers' specifications and create descriptions of the work that's planned. They work on new projects as well as rehabilitation of existing facilities. They not only engineer and design projects, using drafting software such as AutoCAD (computer-aided design), but also estimate the costs. Familiarity with blueprints, spreadsheets, databases, Microsoft Office, and other related programs is often required. Hydroelectric engineers perform engineering studies and computations and conduct field analysis, inspection, and performance tests. They also evaluate and monitor hydroelectric equipment performance, and compile findings in reports. They may gather data and present their findings to management, or direct others in conducting this work. They map out the testing, operating, and maintenance plans for the project; create project schedules and technical specifications; and oversee installation and construction, making sure company standards and procedures are followed. They may also be involved in business development efforts by preparing proposals and bids, and participating in pitch meetings.

   Hydroelectric engineers work with other engineers, technicians, and craftsmen on such hydroelectric machinery as turbines, pumps, fans, boilers, heat exchangers, tanks, piping, scrubbers, compressors, scrubbers, and material handling equipment. Civil hydroelectric engineers analyze and design structures such as concrete dams, pipelines, levees, canals, powerhouses, intakes, spillways, and foundations.

   Some engineers are tasked with covering multiple sites in their work. For example, an Internet advertisement for a substation design engineer listed the job location as: Madison, Wisconsin; Tucson, Arizona; and Billings, Montana. In addition to being able to travel, responsibilities for this job included performing conceptual and detailed designs of high-voltage utility substations; being well versed in SCADA (Supervisory Control and Data Acquisition, a system that collects data from various sensors at a factory, plant, or other remote location and forwards that data to a central computer); ability to create physical layouts of substations; and selecting substation equipment and writing up specifications.

REQUIREMENTS
High School
Classes in algebra, trigonometry, physics, chemistry, geology, computers, and English are a good starting ground for this career.

Knowledge of other languages is also helpful as hydroelectric engineers travel, and sometimes relocate, to other countries for work.

James R. Mahar, PE
Operations Project Manager, Bonneville Lock and Dam, Willamette Falls Lock, Oregon

James R. Mahar, PE (licensed professional engineer), is operations project manager of Bonneville Lock and Dam, and Willamette Falls Lock, Oregon. Providing electrical power since 1938, the Belleville Lock and Dam, situated in the Columbia River Gorge National Scenic Area, links cities in two states: Portland, Oregon, and Vancouver, Washington. The U.S. Army Corps of Engineers operates and maintains Bonneville Lock and Dam for hydropower production, fish and wildlife protection, recreation, and navigation. The Willamette Falls Lock opened in 1873 and was created as a navigation route around the falls. The Corps of Engineers purchased it in 1915.

Mahar is a licensed professional engineer in Oregon and Washington, and is also licensed as a management professional. He started working in engineering 25 years ago and has been with Belleville Dam since 2002, where he manages 165 employees. He was recently promoted to the position of Chief of Operations for Portland District, Corps of Engineers. Since August 1, 2009, Mahar has managed about 600 people and is responsible for additional facilities. His background is in electrical engineering, and he first became interested in the field when his school introduced a new program to the curriculum: electricity for high school students.

What's it take to be an engineer? "You have to be a person," Mahar says. He speaks frequently at high schools and career days, and when he kicks off with this definition, people start paying attention. He believes *everybody* has the skill to be an engineer. Using "scientific knowledge to solve practical problems," and having an aptitude for math are important, no doubt, but the core of success is hard work and dedication.

Postsecondary Training

Hydroelectric engineers usually have a bachelor of science in engineering. Many schools are starting to offer programs in renewable energy, so be sure to explore the Web sites of those you're considering to see if they offer hydroelectricity programs. Electrical engineering students take course work in communications systems, power systems, robotics, and computers. Depending upon their specialty, hydroelectric engineers may study civil engineering, urban planning and architecture, hydrology and water resources engineering, and conservancy. Course work in environmental policy is also recommended.

Certification or Licensing

Engineers who provide service to the general public are required to have a professional engineer (PE) license to practice in the United States. Requirements for the PE license include having a degree from an ABET-accredited engineering program, four years of related work experience, and passing a state examination. Most electrical, mechanical, civil, and chemical engineers have PE licenses.

Certification is not required but can help engineers to advance to more senior positions. Organizations such as the American Academy of Environmental Engineers (http://www.aaee.net) and the Institute of Electrical and Electronics Engineers (http://www.ieeeusa.org) offer certification programs to professional engineers.

Other Requirements

Detail-oriented, problem-solving, analytical, creative people fare best in this type of work. Passion for the environment and the drive to keep up with trends and discoveries in the renewable energy field are important characteristics of hydroelectric engineers. As with most other engineering specialties, the ability to communicate clearly (both in writing and orally), the flexibility to work with diverse teams of people, and strong organizational and project-management skills are key elements to job success and enjoyment.

EXPLORING

The best way to gain a true appreciation for hydroelectricity is to see it in action. If you're lucky enough to live near a dam, go take

Did You Know?

Hydroelectric plants come in a variety of sizes: from large and medium, to small, mini, and micro. Most of the world's hydroelectricity comes from large-scale hydro plants, which generate thousands of megawatts of energy. Small-scale plants generally produce up to 30 megawatts of electricity and are usually installed for flood control or irrigation projects. They are also used to redevelop electric power production at sites that once used waterwheels for milling. Mini-hydros usually provide less than 1,000 kilowatts of power. Micro-hydros provide up to 100 kilowatts of power, and are used to power single families, small communities, or small enterprises.

a tour. Ask your school if they'll organize a group trip, or see if your family will take a drive. You can also learn more by exploring the Web sites of the Niagara Power Plant (http://www.nypa.gov/facilities/niaga ra.htm) and Bonneville Lock and Dam (http://www.nwp.usace.army.mil/op/b/home.asp). You can also see fish swimming through the "ladder" (a structured passageway, also known as fishway or fish steps, to help fish migrate) by going to the Fish Count & Camera section of the Bonneville site.

The Water Science for Schools section of the U.S. Geological Survey's Web site offers great information about how hydroelectricity works (http://ga.water.usgs.gov/edu/wuhy.html). And you can see firsthand examples of dams and hydroelectric projects by watching videos on the How Stuff Works Web site (http://videos.howstuffworks.com/discovery/30198-really-big-things-hydroelectric-power-video.htm).

EMPLOYERS

Hydroelectric engineers work for federal and state governments. They may work for companies providing engineering consulting services, teaming up with other engineers, scientists, technologists, and technicians on projects, or they may provide these same services through companies they own and manage themselves. Power companies hire hydrologists to help develop new facilities and to renovate and maintain existing operations.

STARTING OUT

Get a solid head start in this field by reading and researching as much as you can about hydroelectric power. Read magazines such as *Scientific American* and *Popular Science*. Also, the Water Education Foundation provides a wide range of information about everything to do with water, including water supply and quality, environmental management and flood restoration, endangered species, urban growth, climate growth, and of course, hydroelectricity and dams. Visit the Web site to learn more about upcoming events and projects (http://www.watereducation.org).

ADVANCEMENT

Hydroelectric engineers can enhance their careers by pursuing advanced degrees and getting professional certification. They can move up to more technical positions and management roles. Those who wish to share their knowledge can offer lecture services to professional associations, and teach at universities and colleges.

EARNINGS

There are currently no specific salary statistics for hydroelectric engineers. The U.S. Department of Labor (DoL) reports that civil engineers earned annual incomes ranging from $48,140 to $115,630 or higher in 2008. Salary.com cites annual salaries for entry-level civil engineers as ranging from $44,484 to $62,526 or higher in 2009.

In 2008 the lowest paid 10 percent of electrical engineers brought home $52,990 per year, the middle 50 percent averaged $82,160, and the highest paid 10 percent averaged $125,810 or more, according to the DoL. Mechanical engineers' annual salaries were slightly lower that same year, ranging from $47,900 to $114,740.

Hydroelectric engineers often receive decent benefits in their work, including profit-sharing plans, bonuses, health and life insurance coverage, disability, paid vacation and sick time, and relocation coverage.

WORK ENVIRONMENT

Engineers work in industrial plants, offices, or laboratories. Hydroelectric engineers spend some time outdoors monitoring operations at dams, power plants, and construction sites. Depending on the job, they may also spend some time traveling to different locations to oversee projects and to attend business meetings and conferences.

OUTLOOK

According to the U.S. Department of Labor, employment of electrical and mechanical engineers is expected to be slower than the average for all occupations through 2016. Things look brighter for civil engineers, who are expected to have faster-than-average employment growth in the next decade. Governments are starting to invest more money into renewable energy research and development, which may result in more hydroelectric power projects.

"Awareness of renewable energy has reached the national level," Jim Mahar says. "And energy policy needs to go in this direction. There's a huge field for environmental engineering: hydropower, wind, solar. In the early days, we studied control systems, power engineering, etc., and it was all done before. This is now an exciting field."

Jobs will be competitive for hydroelectric engineers. Those with solid academic records, practical work experience on successful projects, and the flexibility to travel and/or relocate, if needed, will have better chances of landing work.

FOR MORE INFORMATION

For industry information and events, visit

Association of Energy Engineers
4025 Pleasantdale Road, Suite 420
Atlanta, GA 30340-4260
Tel: 770-447-5083
http://www.aeecenter.org

Learn more about renewable energy by visiting

Energy Efficiency and Renewable Energy
U.S. Department of Energy
http://www.eere.energy.gov

Read about sustainability issues and find out about membership at the following Web sites:

International Hydropower Association
Nine Sutton Court Road
Sutton, London SM1 4SZ
United Kingdom

Email: iha@hydropower.org
http://hydropower.org

National Hydropower Association
25 Massachusetts Avenue, NW, Suite 450
Washington, DC 20001-7405
Tel: 202-682-1700
Email: help@hydro.org
http://hydro.org

Find publications, workshops, and programs for students at this organization's Web site.

Water Education Foundation
717 K Street, Suite 317
Sacramento, CA 95814-3477
Tel: 916-444-6240
http://www.watereducation.org

Nuclear Engineers

 QUICK FACTS

School Subjects
Mathematics
Physics

Personal Skills
Communication/ideas
Technical/scientific

Work Environment
Primarily indoors
Primarily one location

Minimum Education Level
Bachelor's degree

Salary Range
$68,300 to $97,080 to $136,880+

Certification or Licensing
Required for certain positions

Outlook
About as fast as the average

OVERVIEW

Nuclear engineers are concerned with accessing, using, and controlling the energy released when the nucleus of an atom is split. The process of splitting atoms, called fission, produces a nuclear reaction, which creates radiation in addition to nuclear energy. Nuclear energy and radiation has many uses. Some engineers design, develop, and operate nuclear power plants, which are used to generate electricity and power navy ships. Others specialize in developing nuclear weapons, medical uses for radioactive materials, and disposal facilities for radioactive waste. There are approximately 15,000 nuclear engineers employed in the United States.

HISTORY

Nuclear engineering as a formal science is quite young, yet part of its theoretical foundation rests with the ancient Greeks. In the 5th century B.C. Greek philosophers postulated that the building blocks of all matter were indestructible elements, which they named *atomos*, meaning "indivisible." This atomic theory was accepted for centuries, until the British chemist and physicist John Dalton revised it in the early 1800s. In the following century, scientific and mathematical experimentation led to the formation of modern atomic and nuclear theory.

Today, it is known that the atom is far from indivisible and that its dense center, the nucleus, can be split to create tremendous energy. The first occurrence of this splitting process was inadvertently induced in 1938 by two German chemists, Otto Hahn and Fritz Strassman. Further studies confirmed this process and established that the fragments resulting from the fission in turn strike the nuclei of other atoms, resulting in a chain reaction that produces constant energy.

The discipline of modern nuclear engineering is traced to 1942, when physicist Enrico Fermi and his colleagues produced the first self-sustained nuclear chain reaction in the first nuclear reactor ever built. In 1950 North Carolina State College offered the first accredited nuclear engineering program. By 1965 nuclear engineering programs had become widely available at universities and colleges throughout the country and worldwide. These programs provided engineers with a background in reactor physics and control, heat transfer, radiation effects, and shielding.

Current applications in the discipline of nuclear engineering include the use of reactors to propel naval vessels and the production of radioisotopes for medical use. Most of the growth in the nuclear industry, however, has focused on the production of electric energy.

Despite the controversy over the risks involved with atomic power, it continues to be used around the world for a variety of purposes. The Nuclear Energy Institute reports that in 2007, nuclear power plants provided approximately 14 percent of the world's electricity production, with 16 countries using nuclear energy to supply 25 percent of their total electricity. There are 30 countries operating 436 nuclear reactors for electricity generation. In the United States 104 nuclear power plants (located in 31 states) supply electricity to about one-fifth of all homes and businesses. Medicine, manufacturing, and agriculture have also benefited from nuclear research. Such use requires the continued development of nuclear waste management. Low-level wastes, which result from power plants as well as hospitals and research facilities, must be reduced in volume, packed in leak-proof containers, and buried, and waste sites must be continually monitored.

Did You Know?

Of the 31 U.S. states with operating nuclear reactors, nuclear energy accounts for the largest percentage of electricity generated in six states. They are:
- Vermont (79.7 percent)
- Connecticut (51.2)
- South Carolina (50.9)
- New Jersey (50.4)
- New Hampshire (41.0)
- New York (31.0)

Source: Nuclear Energy Institute

THE JOB

Nuclear engineers are involved in various aspects of the generation, use, and maintenance of nuclear energy and the safe disposal of its waste. Nuclear engineers work on research and development, design, fuel management, safety analysis, operation and testing, sales, and education. Their contributions affect consumer and industrial power supplies, medical technology, the food industry, and other industries.

Nuclear engineering is dominated by the power industry. Some engineers work for companies that manufacture reactors. They research, develop, design, manufacture, and install parts used in these facilities, such as core supports, reflectors, thermal shields, biological shields, instrumentation, and safety and control systems.

Those who are responsible for the maintenance of power plants must monitor operations efficiently and guarantee that facilities meet safety standards. Nuclear energy activities in the United States are closely supervised and regulated by government and independent agencies, especially the Nuclear Regulatory Commission (NRC). The NRC oversees the use of nuclear materials by electric utility companies throughout the United States. NRC employees are responsible for ensuring the safety of nongovernment nuclear materials and facilities and for making sure that related operations do not adversely affect public health or the environment. Nuclear engineers who work for regulatory agencies are responsible for

A nuclear engineer demonstrates some of the radiation-detection devices that will be used during a spacecraft mission launch. Even though there is still controversy regarding the risks of utilizing nuclear energy, it continues to be used around the world for multiple purposes. Recent growth in the field has focused on the production of electricity. *AP Photo/Terry Renna*

setting the standards that all organizations involved with nuclear energy must follow. They issue licenses, establish rules, implement safety research, perform risk analyses, conduct on-site inspections, and pursue research. The NRC is one of the main regulatory agencies employing nuclear engineers.

Many nuclear engineers work directly with public electric utility companies. Tasks are diverse, and teams of engineers are responsible for supervising construction and operation, analyzing safety, managing fuel, assessing environmental impact, training personnel, managing the plant, storing spent fuel, managing waste, and analyzing economic factors.

Some engineers working for nuclear power plants focus on the quality of the water supply. Their plants extract salt from water, and engineers develop new methods and designs for such desalinization systems.

The food supply also benefits from the work of nuclear engineers. Nuclear energy is used for pasteurization and sterilization, insect and pest control, and fertilizer production. Furthermore, nuclear engineers conduct genetic research on improving various food strains and their resistance to harmful elements.

Nuclear engineers in the medical field design and construct equipment for diagnosing and treating illnesses and disease. They perform research on radioisotopes, which are produced by nuclear reactions. Radioisotopes are used in heart pacemakers, in X-ray equipment, and for sterilizing medical instruments. According to the Nuclear Energy Institute, as of April 2009, nuclear medical imaging procedures had increased seven fold in the last 25 years.

Numerous other jobs are performed by nuclear engineers. *Nuclear health physicists*, *nuclear criticality safety engineers*, and *radiation protection technicians* conduct research and training programs designed to protect plant and laboratory employees against radiation hazards. *Nuclear fuels research engineers* and *nuclear fuels reclamation engineers* work with reprocessing systems for atomic fuels. Accelerator operators coordinate the operation of equipment used in experiments on subatomic particles, and scanners work with photographs, produced by particle detectors, of atomic collisions.

REQUIREMENTS

High School

If you are interested in becoming a professional engineer, you must begin preparing yourself in high school. Take honors-level courses in mathematics and the sciences. Specifically, you should complete courses in algebra, geometry, trigonometry, calculus, chemistry, physics, and biology. Take English, social studies, and a foreign language (many published technical papers that are required reading in later years are written in German or French). Also, be sure to keep your computer skills up to date by taking computer science classes.

Postsecondary Training

Professional engineers must have at least a bachelor's degree. You should attend a four-year college or university that is approved by the Accreditation Board for Engineering and Technology. In a nuclear engineering program, you will focus on subjects similar to those studied in high school but at a more advanced level. Courses also include engineering sciences and atomic and nuclear physics.

These subjects will prepare you for analyzing and designing nuclear systems and understanding how they operate. You will learn and comprehend what is involved in the interaction between radiation and matter; radiation measurements; the production and use of radioisotopes; reactor physics and engineering; and fusion reactions. The subject of safety will be emphasized, particularly with regard to handling radiation sources and implementing nuclear systems.

You must have a master's or doctoral degree for most jobs in research and higher education, and for supervisory and administrative positions. A graduate degree in nuclear engineering is recommended because this level of education will help you obtain the skills required for advanced specialization in the field. Many institutions that offer advanced degrees have nuclear reactors and well-equipped laboratories for teaching and research. You can obtain information about these schools by contacting the U.S. Department of Energy (http://www.energy.gov).

Certification or Licensing

A professional engineer (PE) license is usually required before obtaining employment on public projects (i.e., work that may affect life, health, or property). Although registration guidelines differ for each state, most states require a degree from an accredited engineering program, four years of work experience in the field, and a minimum grade on a state exam.

Other Requirements

Nuclear engineers will encounter two unique concerns. First, exposure to high levels of radiation may be hazardous; thus, engineers must always follow safety measures. Those working near radioactive materials must adhere to strict precautions outlined by regulatory standards. In addition, female engineers who are pregnant may not be allowed to work in certain areas or perform certain duties because of the potential harm to the human fetus from radiation.

Finally, nuclear engineers must be prepared for a lifetime of continuing education. Because nuclear engineering is founded in the fundamental theories of physics, and the notions of atomic and nuclear theory are difficult to conceptualize except through mathematics, an aptitude for physics, mathematics, and chemistry is indispensable.

EXPLORING

If you are interested in becoming an engineer, you can join science clubs such as the Junior Engineering Technical Society (JETS), which has a chapter in almost every state. Science clubs provide

Average Annual Earnings for Nuclear Engineers by Industry, 2008

Architectural and engineering services: $109,280
Scientific research and development services: $107,120
Federal government: $92,630
Power generation and supply: $84,450

Source: U.S. Department of Labor

the opportunity to work with others, design engineering projects, and participate in career exploration. The *Pre-Engineering Times* (http://www.jets.org/publications/petimes.cfm) will introduce you to engineering careers and a wide variety of engineering-related resources. If you are a more advanced student, you may want to read materials published by the American Nuclear Society (http://www.ans.org).

EMPLOYERS

Nuclear engineers work in a variety of settings. According to the U.S. Department of Labor, about 30 percent of the approximately 15,000 nuclear engineers employed in the United States work in research and development in the physical, engineering, and life sciences. Many also work for electrical power companies (such as Union Electric), reactor manufacturers (e.g., General Electric, Westinghouse), architect-engineering firms and consulting firms, national laboratories, and the federal government. Of those who work for the federal government, many are civilian employees of the navy, and most of the rest work for the U.S. Department of Energy (DOE).

STARTING OUT

Most students begin their job search while still in college, collecting advice from job counselors and their schools' career services centers and using organizations and Web sites to find open positions. For example, the Society of Women Engineers (SWE) offers members the opportunity to post their resumes or find job matches through its Web site. Networking with those already employed in the field is an excellent way to find out about job openings. Networking opportunities are available during meetings of professional organizations, such as the SWE's annual national conference.

As with other engineering disciplines, a hierarchy of workers exists, with the chief engineer having overall authority over managers and project engineers. This is true whether you are working in research, design, production, sales, or teaching. After gaining a certain amount of experience, engineers may apply for positions in supervision and management.

ADVANCEMENT

Because the nuclear engineering field is so young, the time is ripe for technological developments, and engineers must therefore keep abreast of new research and technology throughout their careers. Advancement for engineers is contingent upon continuing education, research activity, and on-the-job expertise.

Advancement may also bring recognition in the form of grants, scholarships, fellowships, and awards. For example, the American Nuclear Society (ANS) has established a Young Members Engineering Achievement Award to recognize outstanding work performed by members. To be eligible for this award, you must be an ANS member, younger than 40 years old, and demonstrate effective application of engineering knowledge that results in a concept, design, analysis method, or product used in nuclear power research and development or in a manufacturing application.

EARNINGS

Nuclear engineers earned a median income of $97,080 in 2008, according to the U.S. Department of Labor. The department also reports that the highest paid 10 percent of nuclear engineers earned more than $136,880, while the lowest paid 10 percent earned less than $68,300 annually. Nuclear engineers working for the federal government had an average salary of $92,630 in 2008. Findings of a 2007 salary survey by the National Association of Colleges and Employers show that those with bachelor's degrees in engineering, including nuclear engineering, averaged starting salaries of $56,587.

Benefits offered depend on the employer but generally include paid vacation and sick days, health insurance, and retirement plans.

WORK ENVIRONMENT

In general, nuclear engineering is a technically demanding and politically volatile field. Those who work daily at power plants perhaps incur the most stress because they are responsible for preventing large-scale accidents involving radiation. Those who work directly with nuclear energy face risks associated with radiation contamination. Engineers handling the disposal of hazardous material also work under stressful conditions because they must take tremendous care to ensure the public's health and safety.

Research, teaching, and design occupations allow engineers to work in laboratories, classrooms, and industrial manufacturing facilities. Many engineers who are not directly involved with the physical maintenance of nuclear facilities spend most of their working hours, an average of 46 hours per week, conducting research. Most work at desks and must have the ability to concentrate on very detailed data for long periods of time, drawing up plans and constructing models of nuclear applications.

OUTLOOK

According to the U.S. Department of Labor, employment growth for nuclear engineers is expected to be about as fast as the average through 2016. Most openings will arise as nuclear engineers transfer to other occupations or leave the labor force. Good opportunities for nuclear engineers should still exist, however, because the small number of nuclear engineering graduates is likely to be in balance with the number of job openings.

The Nuclear Energy Institute (NEI) reported that in 2008, approximately 74 percent of 1,000 Americans surveyed support the use of nuclear energy. Additionally, 69 percent support the construction of new nuclear power plants in the future. The Nuclear Power 2010 program is a joint effort between the federal government and private industry, initiated in 2002, to identify sites for new advanced nuclear power plants and begin construction by 2010. According to the NEI, 18 new facilities applied for construction/operating licenses between 2007 and 2009. Even if new plants are not constructed, nuclear engineers will be needed to operate existing plants. They will also continue to be needed to work in defense-related areas, to develop nuclear-related medical technology, and to monitor and improve waste management and safety standards.

FOR MORE INFORMATION

For information on scholarships, education, and careers, contact
American Nuclear Society
555 North Kensington Avenue
LaGrange Park, IL 60526-5592
Tel: 800-323-3044
http://www.ans.org

For information on student membership, contact
Junior Engineering Technical Society Inc.
1420 King Street, Suite 405
Alexandria, VA 22314-2750
Tel: 703-548-5387
Email: info@jets.org
http://www.jets.org

For a wide variety of career and industry information, contact
Nuclear Energy Institute
1776 I Street, NW, Suite 400
Washington, DC 20006-3708
Tel: 202-739-8000
http://www.nei.org

For career guidance and scholarship information, contact
Society of Women Engineers
230 East Ohio Street, Suite 400
Chicago, IL 60611-3265
Tel: 312-596-5223
Email: hq@swe.org
http://societyofwomenengineers.swe.org/

For information on careers and nuclear power, contact
U.S. Department of Energy
1000 Independence Avenue, SW
Washington, DC 20585-0001
Tel: 800-342-5363
http://www.energy.gov

Nuclear Reactor Operators and Technicians

QUICK FACTS

School Subjects
Mathematics
Physics

Personal Skills
Following instructions
Technical/scientific

Work Environment
Primarily indoors
Primarily one location

Minimum Education Level
Some postsecondary training

Salary Range
$40,310 to $73,320 to $96,480

Certification or Licensing
Required

Outlook
Little or no change (operators)
About as fast as the average (technicians)

OVERVIEW

Licensed *nuclear reactor operators* work in nuclear power plant control rooms, where they monitor instruments that record the performance of every pump, compressor, and other treatment system in the reactor unit. Nuclear power plants must have operators on duty at all times. In addition to monitoring the instruments in the control room, the nuclear reactor operator runs periodic tests on equipment at the station. *Nuclear reactor operator technicians* are in training to become operators; they study nuclear science theory and learn to perform reactor operation and control activities. They work under the supervision of licensed nuclear reactor operators, and later they work as beginning operators. *Senior operators*, or *senior reactor operators*, have further training and experience and oversee the activities of nuclear reactor operators and technicians.

HISTORY

The potential for nuclear power generation was first demonstrated in 1942, when a group of scientists led by Enrico Fermi conducted the first controlled nuclear chain reaction in a nuclear reactor located under the football stands on Stagg Field at the University of Chicago. After World War II, research continued on peacetime uses of controlled atomic energy. In 1948 researchers increasingly emphasized the design of nuclear power reactors to generate electricity.

By late 1963, the technology for these nuclear reactors was ready for commercial use, and the first nuclear power plants were constructed. Their successful operation and the low cost of the electric power they generated were promising. Further development of technology continued, and the construction of several additional nuclear power plants began.

Since then, the field has learned a great deal about the design and safe operation of nuclear-fueled electric power plants. Quality assurance and control procedures have been developed to ensure

Did You Know?

- Thirty-one states have operating nuclear reactors.
- The oldest nuclear plant operating in the United States is Oyster Creek in New Jersey. It received its license in 1969.
- The newest nuclear plant is Watts Bar 1 in Tennessee. It began operating in 1996.
- The Palo Verde, Arizona, nuclear plant has three reactors and is the largest such facility in the United States. It is the 12th-largest nuclear power plant in the world.
- In 2009 nuclear power accounted for about 20 percent of the United States' total electricity generation.

Source: Nuclear Energy Institute

that every step of a plant's construction and operation meets the necessary safety requirements.

Specific procedures are in place to protect against radiation, and special technicians work in each plant to ensure the least possible risk of radiation exposure to workers. Studies show that the safest operation of nuclear plants is directly attributable to carefully selected and thoroughly trained nuclear reactor operators. Since 1963, thousands of people have been trained and licensed by the federal government to work as nuclear reactor operators.

THE JOB

Technicians are trained to learn and perform all the duties expected of licensed operators. Almost all the skills and knowledge, however, are learned outside of the reactor control room.

The nuclear reactor is like an engine providing power, in the form of hot steam, to run the entire nuclear power plant. Nuclear reactor operators are the nuclear station's driver in the sense that they control all the machines used to generate power at the station. Working under the direction of a plant manager, the nuclear reactor operator is responsible for the continuous and safe operation of a reactor. Although most nuclear power plants contain more than one nuclear reactor unit, each nuclear reactor operator is responsible for only one of the units.

From the standpoint of safety and uninterrupted operation, the nuclear reactor operator holds the most critical job in the plant. The operator's performance is considered so essential that any shutdown of an average 1,000-megawatt plant, whether due to an accident or operating error, can result in a minimum loss of the cost of the operator's salary for 10 years.

Licensed nuclear reactor operators work in the station control room, monitoring meters and gauges. They read and interpret instruments that record the performance of every valve, pump, compressor, switch, and water treatment system in the reactor unit. When necessary, they make adjustments to the fission rate, pressure, water temperature, and flow rate of the various pieces of equipment to ensure safe and efficient operation.

During each 24-hour period, operators make rounds four times. (They usually work in eight-hour or 12-hour shifts.) This task involves reviewing the unit's control board and writing down the parameters of the instruments. Each hour, a computer generates a reading indicating the amount of power the unit is generating.

In addition to monitoring the instruments in the control room, the nuclear reactor operator runs periodic tests, including pressure readings, flow readings, and vibration analyses on each piece of equipment. The operator must also perform logic testing on the electrical components in order to check the built-in safeguards.

Every 12 to 18 months, the nuclear reactor operator must also refuel the reactor unit, a procedure that is sometimes called an outage. During the refueling, the turbine is brought offline, or shut down. After it cools and depressurizes, the unit is opened, and any repairs, testing, and preventive maintenance are taken care of. Depleted nuclear fuel is exchanged for new fuel. The unit is then repressurized, reheated, and brought back online (or restarted).

Auxiliary equipment operators normally work at the site of the equipment. Their work can include anything from turning a valve to bringing a piece of equipment in and out of service. All of their requests for action on any machine must be approved by the nuclear reactor operator.

Precise operation is required in nuclear power plants to be sure that radiation does not contaminate the equipment, the operating personnel, or the nearby population and environment. The most serious danger is the release of large amounts of atomic radiation into the atmosphere. Operating personnel are directly involved in the prevention of reactor accidents and in the containment of radioactivity in the event of an accident.

Nuclear reactor operators always begin their employment as technicians. In this capacity, they gain plant experience and technical knowledge at a functioning nuclear power plant. The technician trains on a simulator (which is built and equipped as an operating reactor control station) and studies the reactor and control room. Technicians can practice operating the reactor and learn what readings the instruments in the simulator give when certain adjustments are made in the reactor control settings. This company-sponsored training is provided to help technicians attain the expertise necessary to obtain an operator's license. Even after obtaining a license, however, beginning operators work under the direction of a shift supervisor, senior operator, or other management personnel.

REQUIREMENTS

Although a college degree is not required, many utilities prefer candidates to have some postsecondary training. More and more nuclear reactor operators have completed at least two years of college, and about 25 percent have a four-year degree. Lack of college experience, however, does not exclude an applicant from being hired. High school graduates are selected based on subjects studied and aptitude test results.

High School
If you wish to enter nuclear technology programs, take courses in algebra, geometry, English composition, blueprint reading, and chemistry and physics with laboratory study. In addition, classes in computer science and beginning electronics will help you prepare for the technology program that follows high school.

Postsecondary Training
In the first year of a nuclear technology program at a technical or community college, you will probably take nuclear technology, radiation physics, applied mathematics, electricity and electronics, technical communications, basic industrial economics, radiation detection and measurement, inorganic chemistry, radiation protection, mathematics, basic mechanics, quality assurance and quality control, principles of process instrumentation, heat transfer and fluid flow, metallurgy, and metal properties.

In the second year, you may be required to take technical writing and reporting, nuclear systems, blueprint reading, mechanical component characteristics and specifications, reactor physics, reactor safety, power plant systems, instrumentation and control of reactors and plant systems, power plant chemistry, reactor operations, reactor auxiliary systems, and industrial organizations and institutions.

Upon completing a technical program, you will continue training once you are employed at a plant. On-the-job training includes learning nuclear science theory; radiation detection; and reactor design, operation, and control. In addition, nuclear reactor operator technicians must learn in detail how the nuclear power plant works. Trainees are assigned to a series of work-learn tasks that take them to all parts of the plant. If trainees have been working in the plant as regular employees, their individual training is planned around what they already know. This kind of training usually takes two to three years and includes simulator practice.

The simulator is an exact replica of the station's real control room. The controls in the simulator are connected to an interactive computer. Working under the supervision of a licensed nuclear reactor operator, trainees experience mock events in the simulator, which teach them how to safely handle emergencies.

During this on-the-job training, technicians learn about nuclear power plant materials, processes, material balances, plant operating equipment, pipe systems, electrical systems, and process control. It is crucial to understand how each activity within the unit affects other instruments or systems. Nuclear reactor operator technicians are given written and oral exams, sometimes as often as once a week. In some companies, technicians are dismissed from their job for failing to pass any one training exam.

Some people in the industry believe that one of the most difficult aspects of becoming a nuclear reactor operator is getting hired. Because electric utilities invest a substantial amount of time and money to train nuclear reactor operators, they are extremely selective when hiring.

The application process entails intensive screening, including identity checks, FBI fingerprint checks, drug and alcohol tests, psychological tests, and credit checks. After passing this initial screening, the applicant takes a range of mathematical and science aptitude tests.

Utility companies recruit most nuclear reactor operator technicians from local high schools and colleges, fossil fuel plants (utilities using nonnuclear sources of energy), and nuclear navy programs. Knowledge of nuclear science and the discipline and professionalism gained from navy experience make veterans excellent candidates. Graduates of two-year programs in nuclear technology also make excellent trainees because they are well versed in nuclear and power plant fundamentals.

The Nuclear Regulatory Commission (NRC) establishes the standards and course content for all nuclear training programs. In addition, each nuclear power plant training program must be accredited by the Institute of Nuclear Power Operations, which was founded in 1979 by industry leaders to promote excellence in nuclear plant operations.

Certification or Licensing

Nuclear reactor operators are required to be licensed; licensure is based on examinations given by the NRC. The licensing process involves having three years of work experience at a power plant (with at least six months in a nuclear plant), and passing several exams, including a physical exam. The first written exam (Generic Fundamentals Examination) covers topics such as reactor theory and thermodynamics. Candidates who pass this exam then take a site-specific exam that includes a written section and an operating test on the power plant's simulator. Candidates who pass these tests receive their licenses. A license is valid for six years and only for the specific power plant for which the candidate applied.

To maintain their licenses, operators must pass an annual practical, or operating, exam and a written requalification exam given by their employers. Requirements for license renewal include certification from the employer that the operator has successfully completed requalification and operating exams and passed a physical.

Other Requirements

Nuclear reactor operators are subject to continuous exams and ongoing training. They must be diligent about keeping their skills and knowledge up to date. A desire for lifelong learning, therefore, is necessary for those doing this work.

Because of the dangerous nature of nuclear energy, the nuclear reactor operator's performance is critical to the safety of other employees, the community, and the environment. Operators must perform their job with a high degree of precision and accuracy. They must be able to remain calm under pressure and maintain sound judgment in emergencies.

Although nuclear reactor operators must frequently perform numerous tasks at once, they must also be able to remain alert during quiet times and handle the monotony of routine readings and tests.

Responding to requests from other personnel, such as the auxiliary operators, is a regular part of the nuclear reactor operator's job. The ability to communicate and work well with other team members and plant personnel is essential.

EXPLORING

High school guidance counselors and advisers at community or technical colleges are good sources of information about a career as a nuclear reactor operator. The librarians in these institutions also may be helpful in directing you to introductory literature on nuclear reactors.

Opportunities for exploring a career as a nuclear reactor operator are limited because nuclear power plants are usually located in places relatively far from schools and have strictly limited visiting policies. Very few commercial or research reactors provide tours for the general public. Many utility companies with nuclear power plants have visitors' centers, however, where tours are scheduled at specified hours. In addition, interested high school students usually can arrange visits to nonnuclear power plants, which allows them to learn about the energy-conversion process common to all steam-powered electric power generation plants.

EMPLOYERS

There are slightly more than 104 commercial nuclear power plants operating in 31 states in the United States, according to the Nuclear Energy Institute. In addition, there are approximately 32 low-power nuclear reactors used for research and training (known as RTRs) at educational and other institutions, according to the NRC. Nuclear reactor operators, naturally, work at nuclear power plants and are employed by utility or energy companies, universities, and other institutions operating these facilities.

STARTING OUT

In recent years, nuclear technology programs have been the best source for hiring nuclear reactor operator technicians. Students are usually interviewed and hired by the nuclear power plant personnel recruiters toward the end of their technical college program and start working in the power plant as trainees after they graduate.

Navy veterans from nuclear programs and employees from other parts of the nuclear power plant may also be good candidates for entering a training program for nuclear reactor operators.

ADVANCEMENT

Many licensed reactor operators progress to the position of senior reactor operator (as they gain experience and undergo further study). To be certified as senior reactor operators (SROs), operators must pass the senior reactor operator exam, which requires a broader and more detailed knowledge of the power plant, plant procedures, and company policies. In some locations, the senior reactor operator may supervise other licensed operators.

SROs may also advance into the positions of field foreman and then control room supervisor or unit supervisor. These are management positions, and supervisors are responsible for an operating crew. Successful supervisors can be promoted to shift engineer or even plant manager.

Licensed nuclear reactor operators and senior reactor operators may also become part of a power plant's education staff or gain employment in a technical or four-year college, company employee training department, or an outside consulting company. Both operators and SROs may work for reactor manufacturers and serve as research and development consultants. They also may teach trainees to use simulators or operating models of the manufacturer's

Learn More about It

Heaberlin, Scott W. *A Case for Nuclear-Generated Energy: (Or Why I Think Nuclear Power Is Cool and Why It Is Important You Think So Too)*. Columbus, Ohio: Battelle Press, 2003.

Herbst, Alan M., and George W. Hopley. *Nuclear Energy Now: Why the Time Has Come for the World's Most Misunderstood Energy Source*. Hoboken, N.J.: Wiley, 2007.

Hewitt, Geoffrey F. *Introduction to Nuclear Power*. 2d ed. New York: Taylor & Francis, 2000.

Lillington, John N. *The Future of Nuclear Power*. San Diego, Calif.: Elsevier Science, 2004.

Mahaffey, James. *Atomic Awakening: A New Look at the History and Future of Nuclear Power*. New York: Pegasus, 2009.

McCarty, Aaron. *Tales from Nuclear Power School*. North Charleston, S.C.: BookSurge Publishing, 2006.

Perin, Constance. *Shouldering Risks: The Culture of Control in the Nuclear Power Industry*. Princeton, N.J.: Princeton University Press, 2006.

Ramsey, Charles B. *Commercial Nuclear Power: Assuring Safety for the Future*. North Charleston, S.C.: BookSurge Publishing, 2006.

reactors. Finally, operators and SROs may work for the NRC, which administers license examinations.

EARNINGS

The beginning salary rate for nuclear reactor operator technicians depends on the technician's knowledge of nuclear science theory and work experience. Graduates of strong nuclear technology programs or former navy nuclear technicians usually earn more than people without this background or training. Salaries also vary among different electric power companies.

According to the U.S. Department of Labor, in 2008 the lowest paid reactor operator technicians earned less than $40,310 per year, with a median of $67,890. Reactor operators earned an annual median income of $73,320 in 2008, according to the U.S. Department of Labor. Salaries ranged from less than $55,730 to more than $96,480 a year.

In addition to a base salary, some operators are paid a premium for working certain shifts and overtime. Standard benefits include insurance, paid holidays, vacations, and retirement benefits.

Employers also pay for the continued formal and on-the-job training of nuclear reactor operators. Of licensed reactor operator staff members, 10 to 20 percent are in formal retraining programs at any one time to renew their operator's licenses or to obtain a senior operator's license.

WORK ENVIRONMENT

Nuclear reactor operator technicians spend their working hours in classrooms and laboratories, learning about every part of the power plant. Toward the end of their training, they work at a reactor control-room simulator or in the control room of an operational reactor unit under the direction of licensed operators.

Operators work in clean, well-lit (but windowless) control rooms. Because nuclear reactor operators spend most of their time in the control room, employers have made great efforts to make it as comfortable as possible. Some control rooms are painted in bright, stimulating colors, and some are kept a little cooler than is standard in most offices. Some utilities have even supplied exercise equipment for their nuclear reactor operators to use during quiet times.

Because nuclear reactors must operate continuously, operators typically work an eight-hour shift and rotate through each of three shifts, taking turns as required. This means operators will work weekends as well as nights some of the time. During their shift, most operators are required to remain in the control room, often eating their lunches at their station. Being in the same environment for eight hours at a time with the same crew members can be stressful.

Although nuclear reactor operators may work at one station of control boards for a long time, they are not allowed to personalize their space because other operators use each station as the shifts rotate.

Although most operators do not wear suits to work, they dress in office attire. Technicians, however, will spend part of their training outside the reactor area. In this environment, appropriate clothing is worn, including hard hats and safety shoes, if necessary.

Operators are shielded from radiation by the concrete outside wall of the reactor containment vessel. If leaks should occur, operators are less subject to exposure than plant personnel who are more directly involved in maintenance and inspection. Nonetheless, technicians wear film badges that darken with radiation exposure. In addition, radiation measurement is carried out in all areas of the plant and plant surroundings according to a regular schedule.

Operators have the added stress of NRC's tough scrutiny. Plant management, the local community, and the national and local press also watch for compliance with regulatory and safety measures.

A career as a nuclear reactor operator offers the opportunity to assume a high degree of responsibility and to be paid while training. People who enjoy using precision instruments and learning about the latest technological developments are likely to find this career appealing. Operators must be able to shoulder a high degree of responsibility and to work well under stressful conditions. They must be emotionally stable and calm at all times, even in emergencies.

OUTLOOK

Questions regarding the safety of nuclear power, the environmental effects of nuclear plants, and the safe disposal of radioactive waste have been of public concern since the occurrence of major accidents at the Three Mile Island (near Harrisburg, Pennsylvania) and Chernobyl (in Ukraine) plants. Nevertheless, the Nuclear Energy Institute reports that more Americans support the use of nuclear energy and feel nuclear energy is important to the country's future energy needs. Additionally, 69 percent support the construction of new nuclear power plants in the future. A joint effort between the federal government and private industry, Nuclear Power 2010, is charged with identifying sites for new, advanced nuclear power plants to begin construction by 2010.

Many unresolved questions remain about environmental effects and waste disposal and reprocessing. In addition, construction and maintenance costs of nuclear power plants have increased rapidly due to changes in the requirements for power plant design and safety. Until these issues are resolved, despite programs such as Nuclear Power 2010, the future of the nuclear industry will remain uncertain. Most new job openings will occur as a result of retirements or transfers to other jobs. According to the U.S. Department of Labor, employment for all power plant operators is expected to show little or no change through 2016. Technicians are expected to fare slightly better, with employment growth about as fast as the average. Opportunities are predicted to be best in defense-related areas, in developing nuclear medical technology, and improving and enforcing waste management and safety standards.

FOR MORE INFORMATION

For information on publications, scholarships, and seminars, contact

American Nuclear Society
555 North Kensington Avenue
La Grange Park, IL 60526-5592
Tel: 708-352-6611
http://www.ans.org

This organization advocates the peaceful use of nuclear technologies. For information on certification, publications, and local chapters, contact

American Society for Nondestructive Testing
PO Box 28518
1711 Arlingate Plaza
Columbus, OH 43228-0518
Tel: 614-274-6003
http://www.asnt.org

For information on the nuclear industry and careers, as well as a list of academic programs in nuclear energy, contact

Nuclear Energy Institute
1776 I Street, NW, Suite 400
Washington, DC 20006-3708
Tel: 202-739-8000
http://www.nei.org

For information on licensing, contact

U.S. Nuclear Regulatory Commission
Washington, DC 20555-0001
Tel: 800-368-5642
Email: opa@nrc.gov
http://www.nrc.gov

Petroleum Engineers

QUICK FACTS

School Subjects
Mathematics
Physics

Personal Skills
Helping/teaching
Technical/scientific

Work Environment
Indoors and outdoors
One location with some travel

Minimum Education Level
Bachelor's degree

Salary Range
$57,820 to $108,020 to $166,400+

Certification or Licensing
Required for certain positions

Outlook
Slower than the average

OVERVIEW

Petroleum engineers apply the principles of geology, physics, and the engineering sciences to the recovery, development, and processing of petroleum. As soon as an exploration team has located an area that could contain oil or gas, petroleum engineers begin their work, which includes determining the best location for drilling new wells, and the economic feasibility of developing them. They are also involved in operating oil and gas facilities, monitoring and forecasting reservoir performance, and utilizing enhanced oil recovery techniques that extend the life of wells. They also make sure that the drilling process has minimal impact on the land, water, air, vegetation, natural habitats, and surrounding communities. There are approximately 17,000 petroleum engineers employed in the United States.

HISTORY

From a broad perspective, the history of petroleum engineering can be traced back hundreds of millions of years to when the remains of plants and animals blended with sand and mud and transformed into rock. It is from this ancient underground rock that petroleum is taken, for the organic matter of the plants and animals decomposed into oil during these millions of years and accumulated into pools deep underground.

In primitive times, people did not know how to drill for oil; instead, they collected the liquid substance after it had seeped to above ground surfaces. Petroleum is known to have been used at that time for caulking ships and for concocting medicines.

Petroleum engineering as we know it today was not established until the mid-1800s, an incredibly long time after the fundamental ingredients of petroleum were deposited within the earth. In 1859 the American Edwin Drake was the first person to pump the so-called rock oil from under the ground, an endeavor that, before its success, was laughed at and considered impossible. Forward-thinking investors, however, had believed in the operation and thought that underground oil could be used as inexpensive fluid for lighting lamps and for lubricating machines (and therefore could make them rich). The drilling of the first well, in Titusville, Pennsylvania (1869), ushered in a new worldwide era: the oil age.

At the turn of the century, petroleum was being distilled into kerosene, lubricants, and wax. Gasoline was considered a useless by-product and was run off into rivers as waste. The invention of the internal combustion engine and the automobile changed all this. By 1915 there were more than half a million cars in the United States, virtually all of them powered by gasoline.

Edwin Drake's drilling operation struck oil 70 feet below the ground. Since that time technological advances have been made, and the professional field of petroleum engineering has been established. Today's operations drill as far down as six miles. Because the United States began to rely so much on oil, the country contributed significantly to creating schools and educational programs in this engineering discipline. The world's first petroleum engineering curriculum was devised in the United States in 1914. Today there are approximately 26 U.S. universities that offer petroleum engineering degrees.

The first schools were concerned mainly with developing effective methods of locating oil sites and with devising efficient machinery for drilling wells. Over the years, as sites have been depleted, engineers have been more concerned with formulating methods for extracting as much oil as possible from each well. Petroleum engineers now focus on issues such as computerized drilling operations; however, because usually only about 40 to 60 percent of each site's oil is extracted, engineers must still deal with designing optimal conditions for maximum oil recovery. They also focus on creating environmentally friendly drilling and extraction processes that comply with today's strict government regulations, such as the National Ambient Air Quality Standards Act.

THE JOB

Petroleum engineer is a rather generalized title that encompasses several specialties, each one playing an important role in ensuring the safe and productive recovery of oil and natural gas. In general, petroleum engineers are involved in the entire process of oil recovery, from preliminary steps, such as analyzing cost factors and environmental risks, to the last stages, such as monitoring the production rate and then repacking the well after it has been depleted.

Petroleum engineering is closely related to the separate engineering discipline of geoscience engineering. Before petroleum engineers can begin work on an oil reservoir, *geological engineers*, along with *geologists* and *geophysicists*, seek out prospective sites, determining the potential for oil. Petroleum engineers develop plans for drilling. Drilling is usually unsuccessful, with eight out of 10 test wells being "dusters" (dry wells) and only one of the remaining two test wells having enough oil to be commercially producible. Engineers are making good use of emerging technologies to more accurately identify potential oil and natural gas deposits and reduce the impact on the environment by unsuccessful drilling. For instance, three-dimensional seismic imaging technology bounces acoustic and electrical vibrations off underground formations to create multidimensional maps showing areas with potential deposits. When a significant amount of oil is discovered, engineers can then begin their work of maximizing oil production at the site.

The development company's *engineering manager* oversees the activities of the various petroleum engineering specialties, including *reservoir engineers*, *drilling engineers*, and *production engineers*.

Reservoir engineers use the data gathered by the previous geoscience studies and estimate the actual amount of oil that will be extracted from the reservoir. It is the reservoir engineers who determine whether the oil will be taken by primary methods (simply pumping the oil from the field) or by enhanced methods (using additional energy such as water pressure to force the oil up). The reservoir engineer is responsible for calculating the cost of the recovery process relative to the expected value of the oil produced and then simulates future performance using sophisticated computer models. Besides performing studies of existing company-owned oil fields, reservoir engineers also evaluate fields the company is thinking of buying.

Drilling engineers work with geologists and drilling contractors to design and supervise drilling operations. They are the engineers involved with the actual drilling of the well. They ask: What will be the most efficient and least environmentally damaging methods for penetrating the earth? It is the responsibility of these workers to supervise the building of the derrick (a platform, constructed over the well, that holds the hoisting devices), choose the equipment, and plan the drilling methods. Drilling engineers must have a thorough understanding of the geological sciences so that they can know, for instance, how much stress to place on the rock being drilled.

Production engineers determine the most efficient methods and equipment to optimize oil and gas production. For example, they establish the proper pumping unit configuration and perform tests to determine well fluid levels and pumping load. They plan field workovers and well stimulation techniques such as secondary and tertiary recovery (for example, injecting steam, water, or a special recovery fluid) to maximize field production.

Various research personnel are involved in this field; some are more specialized than others. They include the *research chief engineer*, who directs studies related to the design of new drilling and production methods, the *oil-well equipment research engineer*, who directs research to design improvements in oil-well machinery and devices, and the *oil-field equipment test engineer*, who conducts experiments to determine the effectiveness and safety of these improvements. These engineers are also involved in helping to develop drilling and extraction processes that reduce the impact to land surfaces and minimize waste. A good example, as referenced by the American Petroleum Institute, includes new directional drilling technologies that allow for wells to be "steered" underground, as opposed to the older approach of drilling straight through. Multilateral drilling, for instance, spares damage to surface areas by drilling down in only one spot, and once beneath ground, radiating out into different directions and at different depths from that single wellbore.

In addition to all of the above, sales personnel play an important part in the petroleum industry. *Oil-well equipment and services sales engineers* sell various types of equipment and devices used in all stages of oil recovery. They provide technical support and service to their clients, including oil companies and drilling contractors.

REQUIREMENTS
High School
In high school you can prepare for college engineering programs by taking courses in mathematics, physics, chemistry, geology, biology, environmental studies, and computer science. Economics, history, and English are also highly recommended because these subjects will improve your communication and management skills. Mechanical drawing and foreign languages are also helpful.

Postsecondary Training
A bachelor's degree in engineering is the minimum requirement. In college, you can follow either a specific petroleum engineering curriculum or a program in a closely related field, such as geophysics or mining engineering. In the United States, there are approximately 26 universities and colleges that offer programs that concentrate on petroleum engineering, many of which are located in California and Texas. The first two years toward the bachelor of science degree involve the study of many of the same subjects taken in high school, only at an advanced level, as well as basic engineering courses. In the junior and senior years, students take more specialized courses: geology, formation evaluation, properties of reservoir rocks and fluids, well drilling, properties of reservoir fluids, petroleum production, and reservoir analysis. Classes in environmental legislation and policy are also beneficial.

Because the technology changes so rapidly, many petroleum engineers continue their education to receive a master's degree and then a doctorate. Petroleum engineers who have earned advanced degrees command higher salaries and often are eligible for better advancement opportunities. Those who work in research and teaching positions are usually required to have these higher credentials.

Students considering an engineering career in the petroleum industry should be aware that the industry uses all kinds of engineers. People with chemical, electrical, geoscience, mechanical, environmental, and other engineering degrees are also employed in this field.

Certification or Licensing
Many jobs, especially public projects, require that the engineer be licensed as a professional engineer (PE). To be licensed, candidates must have a degree from an engineering program accredited by the Accreditation Board for Engineering and Technology. Additional requirements for obtaining the license vary from state to state, but

all applicants must take an exam and have several years of related experience on the job or in teaching.

Other Requirements

Students thinking about this career should enjoy science and math. You need to be a creative problem-solver who likes to come up with new ways to get things done and try them out. You need to be curious, wanting to know why and how things are done. You also need to be a logical thinker with a capacity for detail, and you must be a good communicator who can work well with others.

EXPLORING

One of the most satisfying ways to explore this occupation is to participate in Junior Engineering Technical Society (JETS) programs. JETS participants enter engineering design and problem-solving contests and learn team development skills, often with an engineering mentor. Science fairs and clubs also offer fun and challenging ways to learn about engineering.

Certain students are able to attend summer programs held at colleges and universities that focus on material not traditionally offered in high school. Usually these programs include recreational activities such as basketball, swimming, and track and field. For example, Worcester Polytechnic Institute offers the Frontiers program, a two-week residential session for high school seniors. For more information, visit http://www.admissions.wpi.edu/Frontiers. The American Indian Science and Engineering Society (AISES) also sponsors two- to six-week mathematics and science camps that are open to American Indian students and held at various college campuses.

You can also learn more about the job by talking with someone who has worked as a petroleum engineer. To locate an experienced engineer, search Internet sites that feature career areas to explore, industry message boards, and mailing lists.

Another good way to explore this type of work is by touring oil fields or corporate sites (contact the public relations department of oil companies for more information), or landing a temporary or summer job in the petroleum industry on a drilling and production crew. Trade journals, high school guidance counselors, the placement office at technical or community colleges, and the associations listed at the end of this article are other useful resources that will help you learn more about the career of petroleum engineer.

EMPLOYERS

Major oil companies as well as smaller oil companies employ petroleum engineers to work in oil exploration and production. Some petroleum engineers work for consulting companies and equipment suppliers. The federal government is also an employer of engineers. In the United States, oil or natural gas is produced in 32 states, with most sites located in Texas, Louisiana, California, and Oklahoma, plus offshore regions. Many other engineers work in other oil-producing areas such as the Arctic Circle, China's Tarim Basin, and the Middle East.

STARTING OUT

The most common and perhaps the most successful way to obtain a petroleum engineering job is to apply for positions through college student career services offices. Oil companies often have recruiters who seek potential graduates while they are in their last year of engineering school.

Applicants are also advised to simply check the job sections of major newspapers and apply directly to companies seeking employees. They should also keep informed of the general national employment outlook in this industry by reading trade and association journals, such as the Society of Petroleum Engineers' *Journal of Petroleum Technology*.

Engineering internships and co-op programs where students attend classes for a portion of the year and then work in an engineering-related job for the remainder of the year allow students to graduate with valuable work experience sought by employers. These students are usually employed full time after graduation at the place where they had their internship or co-op job.

As in most engineering professions, entry-level petroleum engineers first work under the supervision of experienced professionals for a number of years. New engineers are typically assigned to a field location where they learn different aspects of field petroleum engineering. Initial responsibilities may include well productivity, reservoir and enhanced recovery studies, production equipment and application design, efficiency analyses, and economic evaluations. Field assignments are followed by other opportunities in regional and headquarters offices.

ADVANCEMENT

After several years working under professional supervision, engineers can begin to move up to higher levels. Workers often formulate a choice of direction during their first years on the job. In the operations division, petroleum engineers can work their way up from the field to district, division, and then operations manager. Some engineers work through various engineering positions from field engineer to staff, then division, and finally chief engineer on a project. Some engineers may advance into top executive management. In any position, however, continued enrollment in educational courses is usually required to keep abreast of technological progress and changes. After about four years of work experience, engineers usually apply for a professional engineer (PE) license so they can be certified to work on a larger number of projects.

Others get their master's or doctoral degree so they can advance to more prestigious research engineering, university-level teaching, or consulting positions. Also, petroleum engineers may transfer to many other occupations, such as economics, environmental management, and groundwater hydrology. Finally, some entrepreneurial-minded workers become independent operators and owners of their own oil companies.

EARNINGS

Petroleum engineers with a bachelor's degree earned average starting salaries of $60,718 in 2007, according to the National Association of Colleges and Employers. A survey by the Society of Petroleum Engineers reports that its worldwide members earned an average salary of $127,200 in 2008. According to the U.S. Department of Labor, in 2008 the median annual salary for petroleum engineers was $108,020. The lowest paid 10 percent earned $57,820 or less while the highest paid 10 percent earned $166,400 or more.

Salary rates tend to reflect the economic health of the petroleum industry as a whole. When the price of oil is high, salaries can be expected to grow; low oil prices often result in stagnant wages.

Fringe benefits for petroleum engineers are good. Most employers provide health and accident insurance, sick pay, retirement plans, profit-sharing plans, and paid vacations. Education benefits are also competitive.

WORK ENVIRONMENT

Petroleum engineers work all over the world: the high seas, remote jungles, vast deserts, plains, and mountain ranges. Petroleum engineers who are assigned to remote foreign locations may be separated from their families for long periods of time or be required to resettle their families when new job assignments arise. Those working overseas may live in company-supplied housing.

Some petroleum engineers, such as drilling engineers, work primarily out in the field at or near drilling sites in all kinds of weather and environments. The work can be dirty and dangerous. Responsibilities such as making reports, conducting studies of data, and analyzing costs are usually tended to in offices either away from the site or in temporary work trailers.

Other engineers work in offices in cities of varying sizes, with only occasional visits to an oil field. Research engineers work in laboratories much of the time, while those who work as professors spend most of their time on campuses. Workers involved in economics, management, consulting, and government service tend to spend their work time exclusively indoors.

OUTLOOK

Employment for petroleum engineers is expected to grow more slowly than the average for all occupations through 2016, according to the U.S. Department of Labor. Despite this prediction, opportunities for petroleum engineers will exist because the number of degrees granted in petroleum engineering is low, leaving more job openings than there are qualified candidates. (According to the Society of Petroleum Engineers, the average age of its members is 52.) Additionally, employment opportunities may improve as a result of the construction of new gas refineries, pipelines, and transmission lines, as well as drilling in areas that were previously off-limits to such development. Stricter environmental regulations will also create need for engineers who are well versed in environmentally friendly drilling and extraction processes and practices.

The challenge for petroleum engineers in the past decade has been to develop technology that lets drilling and production be economically feasible even in the face of low oil prices. For example, engineers had to rethink how they worked in deep water. They used to believe deep wells would collapse if too much oil was pumped out at once. But the high costs of working in deep water plus low oil prices made low volumes uneconomical. So engineers learned how

Top-Paying Industries for Petroleum Engineers and Annual Median Wages, 2008

- Oil and gas extraction: $127,520
- Petroleum products manufacturing: $124,910
- Architectural and engineering services: $119,330
- Management of companies and enterprises: $116,790
- Support activity for mining: $103,060

Source: U.S. Bureau of Labor Statistics

to boost oil flow by slowly increasing the quantities wells pumped by improving valves, pipes, and other equipment used. Engineers have also cut the cost of deep-water oil and gas production in the Gulf of Mexico, predicted to be one of the most significant exploration hot spots in the world for the next decade, by placing wellheads on the ocean floor instead of on above-sea production platforms.

Cost-effective technology that permits new drilling and increases production, with minimal impact on the environment and surrounding communities, will continue to be essential in the profitability of the oil industry. As long as research and development in the field continues to deliver drilling and extraction equipment and techniques that streamline the petroleum business and comply with environmental regulations, petroleum engineers will continue to have a vital role to play.

FOR MORE INFORMATION

For information on careers in petroleum geology, contact
American Association of Petroleum Geologists
1444 South Boulder
Tulsa, OK 74119-3604
Tel: 800-364-2274
http://www.aapg.org

For information on summer programs, contact
American Indian Science and Engineering Society
PO Box 9828

Albuquerque, NM 87119-9828
Tel: 505-765-1052
Email: info@aises.org
http://www.aises.org

For general information on the petroleum industry, contact
American Petroleum Institute
1220 L Street, NW
Washington, DC 20005-4070
Tel: 202-682-8000
http://www.api.org

For information about JETS programs, products, and engineering career brochures (all disciplines), contact
Junior Engineering Technical Society (JETS)
1420 King Street, Suite 405
Alexandria, VA 22314-2750
Tel: 703-548-5387
Email: info@jets.org
http://www.jets.org

For a petroleum engineering career brochure, a list of petroleum engineering schools, and scholarship information, contact
Society of Petroleum Engineers
222 Palisades Creek Drive
Richardson, TX 75080-2040
Tel: 972-952-9393
Email: spedal@spe.org
http://www.spe.org

For a Frontiers program brochure and application, contact
Worcester Polytechnic Institute
100 Institute Road
Worcester, MA 01609-2280
Tel: 508-831-5286
Email: frontiers@wpi.edu
http://www.admissions.wpi.edu/Frontiers

Petroleum Technicians

QUICK FACTS

School Subjects
Mathematics
Physics

Personal Skills
Helping/teaching
Technical/scientific

Work Environment
Indoors and outdoors
Primarily multiple locations

Minimum Education Level
High school diploma

Salary Range
$26,630 to $53,360 to $97,380+

Certification or Licensing
None available

Outlook
About as fast as the average

OVERVIEW

Petroleum technicians work in a wide variety of specialties. Many kinds of *drilling technicians* drill for petroleum from the earth and beneath the ocean. *Loggers* analyze rock cuttings from drilling and measure characteristics of rock layers. Various types of *production technicians* "complete" wells (prepare wells for production), collect petroleum from producing wells, and control production. *Engineering technicians* provide technical assistance and help improve drilling technology that maximizes field production and minimizes environmental disturbance. *Maintenance technicians* keep machinery and equipment running smoothly. There are approximately 12,000 petroleum and geological technicians employed in the United States.

HISTORY

In the 1950s and 1960s the oil industry was relatively stable. Oil was cheap and in high demand. The international oil market was dominated by the "seven sisters"—Shell, Esso, BP, Gulf, Chevron, Texaco, and Mobil. By the end of the 1960s, though, Middle Eastern countries became more dominant. Many nationalized the major oil companies' operations or negotiated to control oil production. To promote and protect their oil production and revenues gained, Iran, Iraq, Kuwait, Saudi Arabia, and Venezuela formed OPEC (the Organization of Petroleum Exporting Countries). The Arab producers' policies during the Arab/Israeli War of 1973–1974 and the Iranian Revolution in 1978 disrupted oil supplies and skyrocketed oil prices, indicating just how powerful OPEC had become.

By the early 1980s, economic recession and energy conservation measures had resulted in lower oil prices. There was and still is worldwide surplus production capacity. OPEC, which expanded membership to countries in the Far East and Africa, tried to impose quotas limiting production, with little success. In 1986 prices, which had once again risen, plummeted.

The events of the 1970s and 1980s significantly altered the nation's attitude toward the price and availability of petroleum products. Domestic oil companies came to realize that foreign sources of oil could easily be lost through regional conflicts or international tensions. The drop in prices during the mid-1980s, however, reinforced the need for domestic producers to continue to find economical oil-producing methods to remain competitive with foreign-produced oil. Today, concerns in the oil industry have expanded to include not only exploring and drilling more efficiently, but also preserving landscapes, protecting wildlife habitats, and minimizing waste.

These developments have fostered great changes in the technology of oil drilling, in the science related to oil exploration, and in the management of existing oil fields. In many abandoned fields, scientists found that they still had nearly as much oil as had been produced from them by older methods. New technology is constantly being developed and used to find economical and environmentally responsible ways of extracting more of this remaining oil economically from old and new fields alike.

The petroleum technician occupation was created to help the industry meet such challenges. Technological changes require scientifically competent technical workers as crew members for well drilling and oil field management. Well-prepared technicians are essential to the oil industry and will continue to be in the future.

THE JOB

Before petroleum technicians can begin work on an oil reservoir, geological exploration teams must hunt for prospective sites. These crews perform seismic surveying, in which sound waves are created and their reflection from underground rocks is recorded by seismographs, to help locate potential sources of oil. Other team members collect and examine geological data or test geological samples to determine petroleum and mineral content. They may also use surveying and mapping instruments and techniques to help locate and map test holes or the results of seismic tests. According to the American Petroleum Institute, new and improved technologies such as satellite-derived gravity and bathymetry, global positioning systems (GPS), and geographic imaging systems (GIS) are enabling more cost-effective oil and gas exploration that causes less damage to the natural environment. Another development in the field is the more efficient technique of "slimhole drilling." Drilling into rocks produces massive amounts of "cuttings" or rubble. The slimhole drill is a smaller drill that produces at least 30 percent fewer cuttings.

Drilling has traditionally been the only way to ultimately prove whether or not there is oil; and usually only 20 percent of drilling operations were successful (meaning, oil was found). Such a small success rate has lead to many environments being disturbed and then abandoned. Directional drilling was developed to prevent unnecessary environmental damage by limiting drilling to a single site, rather than drilling and disturbing land surfaces at multiple sites. For example, multilateral drilling is a technique used to drill down in one site and then radiate out beneath the ground in different directions to search for oil. The focus is on going horizontal, as opposed to the past focus of simply drilling vertically—which, in the past, could sometimes be up to five miles deep.

Drilling for oil is a highly skilled operation involving many kinds of technicians, including *rotary drillers, derrick operators, engine operators,* and *tool pushers*.

In the most common type of drilling, a drill bit with metal or diamond teeth is suspended on a drilling string consisting of 30-foot pipes joined together. The string is added to as the bit goes deeper. The bit is turned either by a rotary mechanism on the drill floor or, increasingly, by a downhole motor. As drilling progresses, the bit gets worn and has to be replaced. The entire drilling string, sometimes weighing more than 100 tons, must be hauled to the surface and dismantled section by section, the bit replaced, then the string reassembled and run back down the well. Known as a "round trip," this operation can take the drilling crew most of a 12-hour shift in a deep well. Until recently, drill strings were mostly manually handled; however, mechanized drill rigs that handle pipe automatically have been introduced to improve safety and efficiency.

The driller directs the crew and is responsible for the machinery operation. The driller watches gauges and works throttles and levers to control the hoisting and rotation speed of the drill pipe and the amount of weight on the bit. Special care is needed as the bit nears oil and gas to avoid a "blowout." Such "gushers" were common in the early days of the oil industry, but today's drilling technicians are trained to prevent them. Drillers also are responsible for recording the type and depth of strata penetrated each day and materials used.

Derrick operators are next in charge of the drilling crew. A derrick is a drilling rig that has a stationary section and a movable boom that's used to raise and lower equipment. The operators work on a platform high up on the derrick and help handle the upper end of the drilling string during placement and removal. They also mix the special drilling "mud" that is pumped down through the pipe to lubricate and cool the bit as well as help control the flow of oil and gas when oil is struck.

Engine operators run engines to supply power for rotary drilling machinery and oversee their maintenance. They may help when the roughnecks (drilling floor and rig workers) pull or add pipe sections.

Tool pushers are in charge of one or more drilling rigs. They oversee erection of the rig, the selection of drill bits, the operation of drilling machinery, and the mixing of drilling mud. They arrange for the delivery of tools, machinery, fuel, water, and other supplies to the drilling site.

One very specialized drilling position is the *oil-well fishing-tool technician*. These technicians analyze conditions at wells where some object, or "fish," has obstructed the borehole. They direct the work of removing the obstacle (lost equipment or broken drill pipes, for example), choosing from a variety of techniques.

During drilling, *mud test technicians*, also called *mud loggers*, use a microscope at a portable laboratory on-site to analyze drill cuttings carried out of the well by the circulating mud for traces of oil. After final depth is reached, technicians called *well loggers* lower measuring devices to the bottom of the well on cable called wireline. Wireline logs examine the electrical, acoustic, and radioactive properties of the rocks and provide information about rock type and porosity, and how much fluid (oil, gas, or water) it contains. These techniques, known as formation evaluation, help the operating company decide whether enough oil exists to warrant continued drilling.

The first well drilled is an exploration well. If oil is discovered, more wells, called appraisal wells, are drilled to establish the limits of the field. Then the field's economic worth and profit are evaluated. If it is judged economically worthwhile and environmentally feasible to develop the field, some of the appraisal wells may be used as production wells. The production phase of the operation deals with bringing the well fluids to the surface and preparing them for their trip through the pipeline to the refinery.

The first step is to complete the well—that is, to perform whatever operations are needed to start the well fluids flowing to the surface—and is performed by *well-servicing technicians*. These technicians use a variety of well-completion methods determined by the oil reservoir's characteristics. Typical tasks include setting and cementing pipe (called production casing) so that the oil can come to the surface without leaking into the upper layers of rock. Well-servicing technicians may later perform maintenance work to improve or maintain the production from a formation already producing oil. These technicians bring in smaller rigs similar to drilling rigs for their work.

After the well has been completed, a structure consisting of control valves, pressure gauges, and chokes (called a Christmas tree because of the way its fittings branch out) is assembled at the top of the well to control the flow of oil and gas. Generally, production crews direct operations for several wells.

Well fluids are often a mixture of oil, gas, and water and must be separated and treated before going into the storage tanks. After separation, *treaters* apply heat, chemicals, electricity, or all three to remove contaminants. They also control well flow when the natural pressure is great enough to force oil from the well without pumping.

Pumpers operate, monitor, and maintain production facilities. They visually inspect well equipment to make sure it's functioning properly. They also detect and perform any routine maintenance needs. They adjust pumping cycle time to optimize production and measure the fluid levels in storage tanks, recording the information each day for entry on weekly gauge reports. Pumpers also advise oil haulers or purchasers when a tank is ready for sale.

Gaugers ensure that other company personnel and purchasers comply with the company's oil measurement and sale policy. They spotcheck oil measurements and resolve any discrepancies. They also check pumpers' equipment for accuracy and arrange for the replacement of malfunctioning gauging equipment.

Once a field has been brought into production, good reservoir management is needed to ensure that as much oil as possible is recovered. *Production engineering technicians* work with the production engineers to plan field workovers and well stimulation techniques such as secondary and tertiary recovery (for example, injecting steam, water, or a special recovery fluid) to maximize field production. *Reservoir engineering technicians* provide technical assistance to reservoir engineers. They prepare spreadsheets for analyses required for economic evaluations and forecasts. They also gather production data and maintain well histories and decline curves on both company-operated and outside-operated wells.

The petroleum industry has a need for other kinds of technicians as well, including *geological technicians, chemical technicians,* and *civil engineering technicians.*

REQUIREMENTS

All petroleum technician jobs require at least a high school diploma, and a few specialties require at least a bachelor's degree.

High School

If you are interested in this field, you should begin preparing in high school by taking math, algebra, geometry, trigonometry, and calculus classes. Biology, geology, environmental studies, and physics are other useful classes. High school courses in drafting, mechanics, or auto shop are also valuable preparation, especially for drilling and production technicians. Computer skills are particularly important for engineering technicians, as are typing and English courses.

Postsecondary Training

As mentioned above, postsecondary training is required for only a few petroleum technician positions. For example, a mud test technician must have at least a bachelor's degree in geology. Although postsecondary training is not usually required for drilling, production, or engineering technicians, these workers can gain familiarity with specified basic processes through special education in technical or community colleges. Postsecondary training can also help entry-level workers compete with experienced workers.

Petroleum technology programs, located primarily at schools in the West and Southwest, are helpful both for newcomers to the field and for those trying to upgrade their job skills. An associate's degree in applied science can be earned by completing a series of technical and education courses.

Petroleum technology programs provide training in drilling operations, fluids, and equipment; production methods; formation evaluation along with the basics of core analysis; and well-completion methods and petroleum property evaluation, including evaluation of production history data and basic theories and techniques of economic analysis. These programs emphasize practical applications in the laboratory, field trips, and summer employment, as available.

Suppliers of the special materials, equipment, or services offer specialized training programs designed for oil company employees.

Other Requirements

Petroleum technicians must be able to work with accuracy and precision; mistakes can be costly or hazardous to the technician and to others in the workplace. You should also be able to work both independently and as part of a team, display manual dexterity, mathematical aptitude, and be willing to work irregular hours.

Much of the work in the petroleum industry involves physical labor and is potentially dangerous. Field technicians must be strong and healthy, enjoy the outdoors in all weather, and be flexible and adaptable about working conditions and hours. Drilling crews may be away from their home base for several days at a time, while technicians on offshore rigs must be able to deal with a restricted environment for several days at a time. Petroleum technicians must also like working with machinery, scientific equipment and instruments, and computers. In addition, petroleum technicians must have good eyesight and hearing and excellent hand, eye, and body coordination.

Some technicians must operate off-road vehicles to transport people, supplies, and equipment to drilling and production sites. Most of this task is learned on the job after formal training is completed.

Some petroleum technicians require additional safety training, including hazardous materials training and first-aid training. In some cases, special physical examinations and drug testing are required. Testing and examinations generally take place after technicians are hired.

EXPLORING

You may want to investigate petroleum technician occupations further by checking your school or public libraries for books on the petroleum industry. Other resources include trade journals, high school guidance counselors, the career services office at technical or community colleges, and the associations and Web sites listed at the end of this article. If you live near an oil field, you may be able to arrange a tour by contacting the public relations department of oil companies or drilling contractors.

Summer and other temporary jobs on drilling and production crews are excellent ways of finding out about this field. Temporary work can provide you with firsthand knowledge of the basics of oil field operations, equipment maintenance, safety, and other aspects of the work. You may also want to consider entering a two-year training program in petroleum technology to learn about the field.

EMPLOYERS

Although drilling for oil and gas is conducted in a large number of states, most workers are concentrated in five states: Colorado, Louisiana, Oklahoma, Wyoming, and Texas. Employers in the crude petroleum and natural gas industry include major oil companies and independent producers. About 37 percent of petroleum and geological technicians work for oil and gas extraction companies. The oil and gas field services industry, which includes drilling contractors, logging companies, and well-servicing contractors, is the other major source of employment.

STARTING OUT

You may enter the field of petroleum drilling or production as a laborer or general helper if you have completed high school. From there, you can work your way up to highly skilled technical jobs, responsibilities, and rewards.

Engineering technicians might start out as *engineering* or *production secretaries* and advance to the position of technician after two to five years of on-the-job experience and demonstrated competency in the use of computers.

Other technicians, such as mud test loggers or well loggers, will need a geology degree first. Upon obtaining your degree, you may start out as an assistant to experienced geologists or petroleum engineers.

Generally speaking, industry recruiters from major companies and employers regularly visit the career services offices of schools with petroleum technology programs and hire technicians before they finish their last year of technical school or college.

Because many graduates have little or no experience with well drilling operations, new technicians work primarily as assistants to the leaders of the operations. They may also help with the semi-skilled or skilled work in order to become familiar with the skills and techniques needed.

It is not uncommon, however, for employers to hire newly graduated technicians and immediately send them to a specialized training program. These programs are designed for oil company employees and usually are offered by the suppliers of the special materials, equipment, or services. After the training period, technicians may be sent anywhere in the world where the company has exploratory drilling or production operations.

ADVANCEMENT

In oil drilling and production, field advancement comes with experience and on-the-job competency. Although a petroleum technology degree is generally not required, it is clearly helpful in today's competitive climate. On a drilling crew, the usual job progression is as follows: from roughneck or rig builder to derrick operator, to rotary driller, to tool pusher, and finally, oil production manager. In production, pumpers and gaugers may later become oil company production foremen or operations foremen; from there, they may proceed to operations management, which oversees an entire district. Managers who begin as technicians gain experience that affords them special skills and judgment.

Self-employment also offers interesting and lucrative opportunities. For example, because many drilling rigs are owned by small, private owners, technicians can become independent owners and operators of drilling rigs. The rewards for successfully operating an independent drill can be very great, especially if the owner discovers new fields and shares in the royalties for production.

Working as a consultant or a technical salesperson can lead to advancement in the petroleum industry. Success is contingent upon an excellent record of field success in oil and gas drilling and production.

In some areas, advancement requires further education. Well loggers who want to analyze logs are required to have at least a bachelor's degree in geology or petroleum engineering, and sometimes they need a master's degree. With additional schooling and a bachelor's degree, an engineering technician can become an engineer. For advanced level engineering, a master's degree is the minimum requirement and a doctorate is typically required. Upper-level researchers also need a doctorate.

During periods of rapid growth in the oil industry, advancement opportunities are plentiful for capable workers. Downsizing in recent years has made advancement more difficult, however, and in many cases technicians, geologists, engineers, and others have accepted positions for which they are overqualified.

EARNINGS

Because of their many work situations and conditions, petroleum technicians' salaries vary widely. Salaries also vary according to geographic location, experience, and education. Petroleum and geological technicians had median annual earnings of $53,360 in 2008, according to the U.S. Department of Labor. Salaries ranged from less than $26,630 to $97,380 or more annually.

In general, technicians working in remote areas and under severe weather conditions usually receive higher rates of pay, as do technicians who work at major oil companies and companies with unions. The five states that paid the highest salaries in 2008 were Alaska, California, Colorado, Mississippi, and Texas.

Fringe benefits are good. Most employers provide health and accident insurance, sick pay, retirement plans, profit-sharing plans, and paid vacations. Education benefits are also competitive.

WORK ENVIRONMENT

Petroleum technicians' workplace conditions vary as widely as their duties. They may work on land or offshore, at drilling sites or in laboratories, in offices or refineries.

Field technicians do their work outdoors, day and night, in all kinds of weather. Drilling and production crews work all over the world, often in swamps, deserts, or in the mountains. The work is rugged and physical, and more dangerous than many other kinds of work. Safety is a big concern. Workers are subject to falls and other accidents on rigs, and blowouts can injure or kill workers if well pressure is not controlled.

Drilling crews often move from place to place because work in a particular field may be completed in a few weeks or months. Technicians who work on production wells usually remain in the same location for long periods. Hours are often long for both groups of workers.

Those working on offshore rigs and platforms can experience strong ocean currents, tides, and storms. Living quarters are usually small, like those on a ship, but they are adequate and comfortable. Workers generally live and work on the drilling platform for days at a time and then get several days off away from the rig, returning to shore by helicopter or crew boat.

Engineering technicians generally work indoors in clean, well-lit offices, although some may also spend part of their time in the field.

From Rigs to Reefs: Environmental Ethics in the Oil Industry

In the early days of oil exploration and extraction, the bottom line was the main focus—get oil, make money. Since then, stricter environmental laws and regulations regarding land and water development, wildlife habitats, and pollution have encouraged, and often forced, oil companies to include an environmental ethic in their business considerations and practices. Naturally, the bottom line is still a driving factor in the oil industry, but the environmental impact of its actions is now a major consideration as well. In addition to developing technologies for more efficient and less wasteful drilling and extraction processes (such as seismic imaging, GPS, and GIS technologies; and multilayer and slimhole drilling), oil companies are teaming up with private environmental groups to take steps to better protect sensitive coastal and marine habitats and wildlife. For example, Bristol Resources and The Nature Conservancy (TNC) have been working together on Shamrock Island in Corpus Christi Bay, Texas, to restore and protect habitats impacted by past oil and gas operations. Phillips Petroleum donated $1 million to the TNC's Alaska Chapter, which the organization planned to use for a conservation-planning project to review critical environmental habitats and to buy land or conservation easements to protect those habitats. The oil industry also created a "rigs-to-reefs" program as part of its environmental ethos. In the Gulf of Mexico, for example, more than 120 decommissioned oil rigs have been converted into submerged, artificial reefs, providing habitats for numerous marine species.

Source: American Petroleum Institute

Regular, 40-hour workweeks are the norm, although some may occasionally work irregular hours.

OUTLOOK

Employment of petroleum technicians is expected to grow about as fast as the average for all occupations through 2016, according to the U.S. Department of Labor (DoL). Companies are continuing to restructure and reduce costs in an effort to conserve more money for exploration and drilling abroad and offshore. The implementation of these measures may mean fewer opportunities for petroleum technicians.

Besides looking for new fields, companies are also expending much effort to boost production in existing fields. New cost-effective technology that permits new drilling and increases production, while reducing damage to land surfaces and surrounding communities, will continue to be important in helping the profitability of the oil industry.

Despite its recent difficulties, the oil industry still plays an important role in the economy and employment. More research is being done to create drilling and extraction equipment and techniques that are more cost-effective and meet environmental standards and regulations. Oil and gas will continue to be primary energy sources for many decades. Most job openings will be due to retirements and job transfers. Technicians with specialized training will have the best employment opportunities. The DoL reports that professional, scientific, and technical services firms will increasingly seek the services of petroleum technicians who can act as consultants regarding environmental policy and federal pollution mandates.

FOR MORE INFORMATION

For information on careers in geology and student chapters, contact
American Association of Petroleum Geologists
1444 South Boulder
Tulsa, OK 74119-3604
Tel: 800-364-2274
http://www.aapg.org

For facts and statistics about the petroleum industry, contact
American Petroleum Institute
1220 L Street, NW

Washington, DC 20005-4070
Tel: 202-682-8000
http://www.api.org

For information about JETS programs, products, and engineering career brochures (all disciplines), contact
Junior Engineering Technical Society (JETS)
1420 King Street, Suite 405
Alexandria, VA 22314-2750
Tel: 703-548-5387
Email: info@jets.org
http://www.jets.org

Learn more about conservation programs and partnerships by visiting
The Nature Conservancy
4245 North Fairfax Drive, Suite 100
Arlington, VA 22203-1606
Tel: 703-841-5300
http://www.nature.org

For a list of petroleum technology schools and careers in petroleum engineering, contact
Society of Petroleum Engineers
222 Palisades Creek Drive
Richardson, TX 75080-2040
Tel: 972-952-9393
Email: spedal@spe.org
http://www.spe.org

For a training catalog listing publications, audiovisuals, and short courses, including correspondence courses, contact
The University of Texas at Austin
Petroleum Extension Service
One University Station, R8100
Austin, TX 78712-1100
Tel: 800-687-4132
http://www.utexas.edu/cee/petex

Power Plant Workers

 QUICK FACTS

School Subjects
Mathematics
Technical/shop

Personal Skills
Mechanical/manipulative
Technical/scientific

Work Environment
Primarily indoors
Primarily one
  location

Minimum Education Level
High school diploma

Salary Range
$38,020 to $58,470 to
  $80,390+

Certification or Licensing
Required for certain positions

Outlook
Little or no change

OVERVIEW

Power plant workers include *power plant operators, power distributors,* and *power dispatchers.* In general, power plant operators control the machinery that generates electricity. Power distributors and power dispatchers oversee the flow of electricity through substations and a network of transmission and distribution lines to individual and commercial consumers. The generators in these power plants may produce electricity by converting energy from a nuclear reactor; burning oil, gas, or coal; or harnessing energy from falling water, the sun, or wind.

HISTORY

The first permanent, commercial electric power-generating plant and distribution network was set up in New York City in 1882 under the supervision of the inventor Thomas Edison. Initially, the purpose of the network was to supply electricity to Manhattan buildings equipped with incandescent light bulbs, which had been developed just a few years earlier by Edison. Despite early problems in transmitting power over distance, the demand for electricity grew rapidly. Plant after plant was built to supply communities with electricity, and by 1900 incandescent lighting was a well-established part of urban life. Other uses of electric power were developed as well, and by about 1910, electric power became common in factories, public transportation systems, businesses, and homes.

Many early power plants generated electricity by harnessing water, or hydro, power. In hydroelectric plants, which are often located at dams on rivers, giant turbines are turned by falling water, and that energy is converted into electricity. Until the 1930s, hydroelectric plants supplied most electric power because hydro plants were less expensive to operate than plants that relied on thermal energy released by burning fuels such as coal. Afterward various technological advances made power generation in thermal plants more economical. Burning fossil fuels (coal, oil, or gas) creates heat, which is used to make steam to turn turbines and generate power. During the last several decades, many plants that use nuclear reactors as heat sources for making steam have been in operation.

Today, energy from all these sources—hydropower, burning fossil fuels, and nuclear reactors—is used to generate electricity. Large electric utility systems may generate power from different sources at multiple sites. Although the essentials of generating, distributing, and utilizing electricity have been known for more than a century, the techniques and the equipment have changed. Over the years the equipment used in power generation and distribution has become much more sophisticated, efficient, and centralized, and the use of electric power exceeds the demand for workers.

THE JOB

Workers in power plants monitor and operate the machinery that generates electric power and sends power out to users in a network of distribution lines. Most employees work for electric utility companies or government agencies that produce power, but there are a small number who work for private companies that make electricity for their own use.

In general, power plant operators who work in plants fueled by coal, oil, or natural gas operate boilers, turbines, generators, and auxiliary equipment such as coal crushers. They also operate switches that control the amount of power created by the various generators and regulate the flow of power to outgoing transmission lines. They

A worker makes adjustments on a heat exchanger/generator unit at a geothermal power plant. *AP Photo/Douglas C. Pizac*

keep track of power demands on the system and respond to changes in demand by turning generators on and off and connecting and disconnecting circuits.

Operators must also watch meters and instruments and make frequent tests of the system to check power flow and voltage. They keep records of the load on the generators, power lines, and other equipment in the system, and they record switching operations and any problems or unusual situations that come up during their shifts.

In older plants, *auxiliary equipment operators* work throughout the plant monitoring specific kinds of equipment, such as pumps, fans, compressors, and condensers.

In newer plants, however, these workers have been mostly replaced by automated controls located in a central control room. *Central control room operators* and their assistants work in these nerve centers. Central control rooms are complex installations with many electronic instruments, meters, gauges, and switches that allow skilled operators to know exactly what is going on with the whole generating system and to quickly pinpoint any trouble that needs repairs or adjustments. In most cases, mechanics and maintenance workers are the ones who repair the equipment.

The electricity generated in power plants is sent through transmission lines to users at the direction of *load dispatchers*. Load dispatcher workrooms are command posts where the power generating and distributing activities are coordinated. Pilot boards in the workrooms are like automated maps that display what is going on throughout the entire distribution system. Dispatchers operate converters, transformers, and circuit breakers based on readings given by monitoring equipment.

By studying factors, such as weather, that affect power use, dispatchers anticipate power needs and tell control room operators how much power will be needed to keep the power supply and demand in balance. If there is a failure in the distribution system, dispatchers redirect the power flow in transmission lines around the problem. They also operate equipment at substations, where the voltage of power in the system is adjusted.

REQUIREMENTS

High School
Most employers prefer to hire high school graduates for positions in this occupational field, and often college-level training is desirable. If you are interested in this field, focus on obtaining a solid background in mathematics and science.

Postsecondary Training
Beginners in this field may start out as helpers or in laborer jobs, or they may begin training for duties in operations, maintenance, or other areas. Those who enter training for operator positions undergo extensive training by their employer, both on the job and in formal classroom settings. The training program is geared toward the particular plant in which they work and usually lasts several years. Even after they are fully qualified as operators or dispatchers, most employees will be required to take continuing education refresher courses.

Certification or Licensing
The Nuclear Regulatory Commission (NRC) regulates power plants that generate electricity using nuclear reactors. The NRC must license operators in nuclear plants, because only NRC-licensed operators are authorized to control any equipment in the plant that affects the operation of the nuclear reactor. Nuclear reactor operators are also required to undertake regular drug testing.

Other Requirements

Although union membership is not necessarily a requirement for employment, many workers in power plants are members of either the International Brotherhood of Electrical Workers or the Utility Workers Union of America. Union members traditionally have been paid better than nonunion members.

EXPLORING

There is little opportunity for part-time or summer work experience in this field. Many power plants (both nuclear and nonnuclear) have visitor centers, however, where you can observe some of the power plant operations and learn about the various processes for converting energy into electricity. You might also locate information on this field at libraries, on the Internet, or by contacting the associations listed at the end of this article.

EMPLOYERS

Employees in the power plant field work in several types of power-generating plants, including those that use natural gas, oil, coal, nuclear, hydro, solar, and wind energies. Because electric utility companies have dominated the energy field, most power plant workers work in electrical utilities. Government agencies that produce

Profile:
Thomas Alva Edison (1847–1931)

Thomas Edison invented the incandescent light bulb and then played a major role in creating an electricity-generating network for the city of New York. The first central electric power plant opened on Pearl Street, New York City, in 1882. Despite start-up and transmission problems, the demand for electricity continued to grow and incandescent lighting was part of most American homes by 1900.

    This prolific inventor was also responsible for inventing many other items, including an electric vote-counting machine, the phonograph, storage batteries, dictating machines, the fluoroscope, and the mimeograph duplicating machine.

power are also employers, as are private companies that make electricity for their own use. Employment opportunities are available in any part of the country, as power plants are scattered nationwide. Approximately 47,000 power plant workers are employed in the United States. Of these, 3,800 are nuclear power plant operators; 8,600 are power distributors and dispatchers; and 35,000 work as operators at other types of power plants.

STARTING OUT

People interested in working in electric power plants can contact local electric utility companies directly. Local offices of utility worker unions may also be sources of information about job opportunities. Leads for specific jobs may be found on employment Web sites such as Earthworks (http://www.earthworks-jobs.com) and Indeed (http://www.indeed.com); in newspaper classified ads; and through the local offices of the state employment service. Graduates of technical training programs can often get help locating jobs from their schools' career services offices.

ADVANCEMENT

After they have completed their training, power plant operators may move into supervisory positions, such as the position of a shift supervisor. Most opportunities for promotion are within the same plant or at other plants owned by the same utility company. With experience and appropriate training, nuclear power plant operators may advance to become senior reactor operators and shift supervisors.

EARNINGS

Salaries for workers in the utilities industry are relatively high, but are based on skills and experience, geographical location, union status, and other factors. Operators in conventional (nonnuclear) power plants earned an average salary of $58,470 in 2008, according to the U.S. Department of Labor. The lowest paid 10 percent of workers earned less than $38,020, while the highest paid 10 percent earned more than $80,390 annually. In that same year, power distributors and dispatchers earned a median salary of $65,890. Nuclear power plant operators averaged $73,320 annually, with salaries ranging from less than $55,730 to more than $96,480. In many cases, employee salaries are supplemented significantly by overtime pay. Overtime often becomes necessary during power outages and severe weather conditions.

Since power plants operate around the clock, employees work multiple shifts, which can last anywhere from four to 12 hours. In general, workers on night shifts are paid higher salaries than workers on day shifts. In addition to their regular earnings, most workers receive benefits, such as paid vacation days, paid sick leave, health insurance, and pension plans.

WORK ENVIRONMENT

Most power plants are clean, well-lighted, and ventilated. Some areas of the plant may be quite noisy. The work of power plant workers is not physically strenuous; workers usually sit or stand in one place as they perform their duties. Risk of falls, burns, and electric shock increases for those who work outside of the control room. Workers must follow strict safety regulations and sometimes wear protective clothing, such as hard hats and safety shoes, to ensure safety and avoid serious accidents.

Electricity is needed 24 hours a day, every day of the year, so power plants must be staffed at all times. Most workers will work some nights, weekends, and holidays, usually on a rotating basis, so that all employees share the stress and fatigue of working the more difficult shifts.

OUTLOOK

Consumer demand for electric power is expected to increase, but power-generating plants will install more automatic control and computerized systems and more efficient equipment, which should limit the growth of operating staffs. The U.S. Department of Labor predicts that employment opportunities for power plant workers will show little or no change through 2016, but opportunities may improve as new electric and nuclear power generating plants are built. (According to the National Energy Institute, there have been 18 applications for new power plant licensing and construction since 2007.) Also, the Obama Administration's proposed energy plan (2009) may generate employment opportunities in the alternative and renewable energy field.

Most job openings will develop when experienced workers retire or leave to go into other occupations. Workers who have solid knowledge of computers and automated equipment will enjoy the best employment prospects. Jobs in electric power plants are seldom affected by ups and downs in the economy, so employees in the field have rather stable jobs.

FOR MORE INFORMATION

For job listings and general information on the power industry, contact

American Public Power Association
1875 Connecticut Avenue, NW, Suite 1200
Washington, DC 20009-5715
Tel: 202-467-2900
http://www.appanet.org

To read the publication Key Facts About the Electric Power Industry, *visit the following Web site:*

Edison Electric Institute
701 Pennsylvania Avenue, NW
Washington, DC 20004-2696
Tel: 202-508-5000
http://www.eei.org

For information on union membership, contact the following organizations:

International Brotherhood of Electrical Workers
900 Seventh Street, NW
Washington, DC 20001-3886
Tel: 202-833-7000
http://www.ibew.org

Utility Workers Union of America
815 16th Street, NW
Washington, DC 20006-4101
Tel: 202-974-8200
http://www.uwua.net

For career and education information, and to learn more about energy issues and policies, contact

Nuclear Energy Institute
1776 I Street, NW, Suite 400
Washington, DC 20006-3708
Tel: 202-739-8000
http://www.nei.org

Renewable Energy Workers

QUICK FACTS

School Subjects
Biology
Chemistry
Mathematics
Physics

Personal Skills
Mechanical/manipulative
Technical/scientific

Work Environment
Indoors and outdoors (technical positions)
Primarily one location (administrative positions)

Minimum Education Level
High school diploma (support positions)
Bachelor's degree (technical and professional positions)

Salary Range
$22,000 to $50,000 to $82,160+

Certification or Licensing
Recommended for most technical positions (certification)
Required for engineering positions (licensing)

Outlook
Faster than the average (bioenergy, geothermal, wind, and solar industries)
Little or no change (hydropower industry)

OVERVIEW

Renewable energy is defined as a clean and unlimited source of power or fuel. This energy is harnessed from different sources such as wind, sunlight (solar), water (hydro), organic matter (biomass), and the earth's internal heat (geothermal). Unlike nonrenewable energy sources like oil, natural gas, coal, or nuclear energy, renewable energy is not based on extracting a limited resource.

The renewable energy industry is actually a vast group of subindustries that offer employment opportunities for people with many different educational backgrounds. *Engineers, scientists, architects, farmers, technicians, operators, mechanics, lawyers, businesspeople, sales workers, human resource and public affairs specialists,* as well as a host of administrative support workers make their living by researching, developing, installing, and promoting renewable energy. The National Renewable Energy Laboratory (NREL), based in Colorado, estimated the renewable energy industries will provide at least 300,000 new jobs for American workers in the next two decades.

HISTORY

Renewable energy resources have been used for centuries. Windmills have long been used to grind grain or pump water. The sun has always been used as a source of heat. In 1839 Alexandre-Edmond Becquerel, an early pioneer in solar energy, discovered the photoelectric effect—the production of electricity from sunlight. The power of water that is stored and released from dams has been used for generating electricity. This type of electricity is known as hydropower electricity. Hot springs and underground reservoirs, products of geothermal energy, have long been used as sources of heat. People have burned trees or other organic matter, known as biomass, for warmth or cooking purposes.

The early technology of harnessing and producing renewable energy as a source of power or fuel was underdeveloped and expensive, however. Because of this, the majority of our power needs have been met using nonrenewable resources such as natural gas or fossil fuels. Our use of fossil fuels has caused our nation to rely heavily on foreign sources to meet demand. Our declining national supply of nonrenewable natural resources, coupled by public awareness of the soaring costs and environmental damage caused by the mining, processing, and use of conventional energy sources, have shed new light on renewable energy sources as a viable solution to our energy needs.

Today, "green" sources of power have earned respect as an important alternative to nonrenewable resources. New research and technology in the past 25 years have enabled self-renewing resources to be harnessed more efficiently and at a lower cost than in the past. Deregulation and a restructuring of the conventional power industries by the Energy Policy Act of 1992 have presented the public with more choices. And the American Recovery and Reinvestment Act of 2009 included more than $60 billion in clean energy investments. Tax incentives at the state and federal level make buying green power more affordable for consumers and for the utility companies. Renewable energy sources are used to produce more than 2 percent of all electricity in the United States, according to the U.S. Department of Energy.

THE JOB

The renewable energy industry can be broken down into the following subindustries: wind, solar, hydropower, geothermal, and bioenergy. A wide variety of career options are available to workers with only a high school diploma and those with advanced degrees. Additionally, many career skills are transferable from one subindustry to another.

Wind

Wind energy has been the fastest-growing energy technology in the world for the past three years, according to the American Wind Energy Association (AWEA). The AWEA reports that more than 1 percent of all U.S. electric power generation was supplied by wind energy in 2007. There are people working in the wind industry in nearly all 50 states.

The wind turbine is the modern, high-tech equivalent of yesterday's windmill. A single wind turbine can harness the wind's energy to generate enough electricity to power a house or small farm. Wind plants, also called wind farms, are a collection of high-powered turbines that can generate electricity for tens of thousands of homes. In order to achieve this capacity, a variety of technical workers are employed in the wind power industry. Electrical, mechanical, and aeronautical engineers design and test the turbines as well as the wind farms. *Meteorologists* help to identify prime locations for new project sites, and may serve as consultants throughout the duration of a project. Skilled construction workers build the farms; *windsmiths*, sometimes called *mechanical or electrical technicians*, operate and maintain the turbines and other equipment on the farm.

Solar

Solar energy is responsible for at least 1 percent of electricity generated by non-hydro renewable energy and approximately .02 percent of the total U.S. electricity, according to the U.S. Department of Energy. Its potential as a major energy source is largely untapped.

There are different ways to turn the sun's energy into a useful power source. The most common technology today uses photovoltaic (PV) cells. When a PV cell is directly struck by sunlight,

>
> ## Why Should You Care about Renewable Energy?
>
> - **Save the environment:** Renewable energy is clean energy. It can be converted with little or no negative impacts on the environment.
> - **Future generations:** Renewable energy sources can never be depleted. This means the technology we set up today to gather power and fuel will benefit generations to come.
> - **The economy:** The renewable energy industry is labor intensive. Many different types of jobs are needed in research, construction, production, and maintenance—many of which are located at or near the project site. Investment dollars are kept close to home, fueling local economies in particular, and that of the United States as a whole.
> - **Energy independence:** Using renewable energy sources helps curb U.S. dependence on fuel from foreign nations.

the materials inside it absorb this light. Simply put, the activity of absorption frees electrons, which then travel through a circuit. Electrons traveling through a circuit produce electricity. Many PV cells can be linked together to produce unlimited amounts of electricity.

The Concentrating Solar Power (CSP) technologies use mirrors to focus sunlight onto a receiver. The receiver collects sunlight as heat, which can be used directly, or generated into electricity. The three CSP methods used are parabolic troughs, power towers, and parabolic dishes. Parabolic troughs can produce solar electricity inexpensively compared to the other methods, and they can generate enough power for large-scale projects. Power towers can also generate power for large-scale projects, while parabolic dishes are used for smaller-scale projects. Using solar collectors and storage tanks, the sun's energy can be used to heat water for swimming pools or buildings. Many schools, hospitals, prisons, and government facilities use solar technology for their water use. A building's design or construction materials can also utilize the sun's energy for its heating and light through passive solar design, water heating, or with electrical PV cells.

Skilled workers are needed for all aspects of solar technology. *Electrical, mechanical, and chemical engineers* work in research and development departments. *Architects,* many of whom specialize in passive solar design and construction, design solar-powered structures. *Technicians, electricians, installers,* and *construction workers* build and maintain solar projects.

Hydropower

Hydropower is the largest and least expensive type of renewable energy in the United States. Hydropower energy is the fourth largest source of electricity generated in the United States; about 10 percent of the total electricity generated in 2003 was from hydropower, according to the U.S. Department of Energy.

Hydropower uses the energy of flowing water to produce electricity. Water is retained in a dam or reservoir. When the water is released, it passes through and spins a turbine. The movement of the turbine in turn spins generators, which produces electricity. In "run of the river" projects, dams are not needed. Canals or pipes divert river water to spin turbines.

Electrical and mechanical engineers and technicians design, construct, and maintain hydropower projects. *Biologists* and other *environmental scientists* assess the effects of hydropower projects on wildlife and the environment. *Fish farmers* develop fish screens and ladders and other migration-assisting devices. *Recreation managers* and *trail planners* manage and preserve the land surrounding the reservoir or dam.

Geothermal

The Geothermal Energy Association states that the United States produces more geothermal energy than any other country, constituting 30 percent of the world total. In 2007 geothermal energy accounted for approximately 4 percent of renewable energy-based electricity consumption in the United States.

Geothermal heat comes from the heat within the earth. Water heated from geothermal energy is tapped from its underground reservoirs and used to heat buildings, grow crops, or melt snow. This direct use of geothermal energy can also be used to generate electricity.

Most water and steam reservoirs are located in the western United States. However, dry rock drilling, a process that drills deeper into the earth's magma, is an innovation that will eventually allow geothermal projects to be undertaken almost anywhere.

Employment opportunities in the geothermal industry are excellent for *geologists, geochemists,* and *geophysicists,* who are needed to research and locate new reservoirs. *Hydraulic engineers, reservoir engineers,* and *drillers* work together to reach and maintain the reservoir's heat supply.

The building of new geothermal projects requires the work of electricians, welders, mechanics, and construction workers. *Drilling workers, machinists,* and *mechanics* also are needed to keep the drilling equipment in good order. *Environmental scientists, chemists,* and other scientists are needed to research and develop new technology to reach other geothermal sources of energy.

Bioenergy

According to the U.S. Department of Energy, bioenergy accounts for 76 percent of electricity generated by non-hydro renewable energy, and 3 percent of total U.S. electricity.

Bioenergy is the energy stored in biomass—organic matter such as trees, straw, or corn. Bioenergy is the second-largest source of renewable energy. It can be used directly, as is the case when we burn wood for cooking or heating purposes. Indirect uses include the production of electricity using wood waste or other biomass waste as a source of power. Another important biomass byproduct is ethanol, which is converted from corn.

Chemists, biochemists, biologists, and *agricultural scientists* work together to find faster and less costly ways to produce bioenergy. *Engineers, construction workers, electricians,* and *technicians* build and maintain bioenergy conversion plants. *Farmers* and *foresters* raise and harvest crops or other sources of biomass. *Truck drivers* transport crops to the conversion plants.

Nontechnical Careers

Within all sectors of the renewable energy industry, nontechnical workers are also needed to perform clerical duties, manage workers, sell, market, and advertise products, maintain records, and educate the public. *Sales and marketing professionals, advertising workers, secretaries, receptionists, customer service representatives, media relations specialists, personnel and human resources specialists, accountants, information technology workers,* and *educators* are just some of the types of nontechnical workers who work in this industry.

REQUIREMENTS

High School

For many jobs in the renewable energy industry, it pays to have a strong background in science and mathematics. For example, earth science, agriculture, and biology classes will be useful if you plan to work in the hydropower industry researching the effects of a new hydropower project on the surrounding vegetation and animal life. Mathematics, earth science, and chemistry classes will be helpful if you plan to work in the geothermal energy industry identifying and harvesting possible sources of geothermal energy from within the earth. Physics classes will be helpful if you plan to work in the wind industry designing windmills and turbine engines to capture and convert wind energy into electricity, or green buildings and homes of the future.

You need not be technically gifted in science and math in order to succeed in the renewable energy industry. Computer classes are useful for workers who run design programs, organize research, and maintain basic office records. Finance, accounting, communications, and English classes will be helpful to anyone who is interested in working in the business end of the industry. Learning a foreign language is also highly useful since a majority of renewable energy companies are located abroad.

Postsecondary Training

Most technical jobs in this industry require at least an associate's or bachelor's degree. Courses of study range from environmental science and mathematics to architecture and meteorology. Many people who are employed in the research and development or technical departments of their respective renewable subindustry have bachelor's or master's degrees in electrical, chemical, or mechanical engineering. Some scientists have graduate degrees in engineering or the sciences (such as biology, physics, or chemistry).

More schools are starting to offer programs in renewable energy, such as Georgia Tech, Purdue University, the University of California (Davis and Irvine), and the University of Wisconsin. San Juan College (http://www.sanjuancollege.edu), located in Farmington, New Mexico, offers an associate's degree or one-year certificate, emphasis in photovoltaic applications, as well as a focus on solar, wind, hydro, and fuel cell applications.

Four-year degrees in liberal arts, business, or other professional degrees are not required, but are recommended for many nontechnical jobs. For example, a *community affairs representative* or *public relations specialist* should have a communications or journalism background.

Certification or Licensing

The Association of Energy Engineers (AEE) offers certification in a variety of specialties. To be considered for certification, a candidate must meet eligibility standards such as a minimum of three years of relevant work experience and membership in a professional organization. Most programs consist of classroom work and examination. One of the most popular certifications is the certified engineering manager (CEM) designation. Many larger utility companies look for the CEM designation when hiring new employees. In fact, according to an AEE survey, over 51 percent of respondents held a CEM or other designation.

Certification and licensing requirements for other jobs in the renewable industry will vary according to the position. Solar panel installers must be certified in order to work on most projects, especially government contracts. Different associations offer certification needs and continuing education training. For example, the Midwest Energy Association offers certification for those working with photovoltaics.

Contractors in the solar industry must apply for certification to ensure their structures are sound and that they meet industry standards. Check the industry trade associations for specifics on project certification.

Most states require engineers to be licensed. There are two levels of licensing for engineers. Professional engineers (PEs) have graduated from an accredited engineering curriculum, have four years of engineering experience, and have passed a written exam. Engineering graduates need not wait until they have four years experience, however, to start the licensure process. Those who pass the Fundamentals of Engineering examination after graduating are called engineers in training (EIT) or engineer interns (EI). The EIT certification usually is valid for 10 years. After acquiring suitable work experience, EITs can take the second examination, the Principles and Practice of Engineering exam, to gain full PE licensure.

Electricians may need licensure depending on the requirements of their job, as well as the industry sector for which they are employed. All states and the District of Columbia require that architects be licensed before contracting to provide architectural services in that particular state. Though many work in the field without licensure, only licensed architects are required to take legal responsibility for all work.

Truck drivers must meet federal requirements and any requirements established by the state where they are based. All drivers must obtain a state commercial driver's license. Truck drivers involved in interstate commerce must meet requirements of the U.S. Department of Transportation.

Other Requirements

It's not absolutely necessary to be a technical genius to do well in this industry. "Much of the technical side can be, and is, taught while on the job," says Katy Mattai, director of a regional energy association. "However, it is important to have an interest in environmental issues. If you don't care about saving our environment, or conserving natural resources, maybe you should reconsider this career choice."

Teamwork is important within all sectors of renewable energy. The ability to work with large groups of people, with varying backgrounds and technical knowledge, is a must.

EXPLORING

Volunteering is one way to explore the renewable energy industry. Katy Mattai discovered this industry after volunteering at a local energy fair. You can find energy fairs or conventions in your area by contacting energy associations. Your duties may consist of handing out brochures or other simple tasks, but you will have the opportunity to learn about the industry and make contacts.

Many professional associations have student chapters or junior clubs. The National Society of Professional Engineers, for example, has local student chapters specifically designed to help high school and college students learn more about careers in engineering. In addition to providing information about different engineering disciplines, student chapters promote contests and offer information on scholarships and internships.

Industry associations also hold many competitions designed to promote their particular renewable energy sector. Visit the student section of the National Renewable Energy Laboratory's (NREL) Web site (http://nrel.gov/learning/student_resources.html) for a list of student programs and competitions held throughout the United States. One such contest is the Junior Solar Sprint held in Colorado for junior high school students. The contest calls for the construction and racing of solar-powered cars. Contestants learn about renewable energy technologies and concepts in a fun, challenging, and exciting setting.

EMPLOYERS

The renewable energy industry is a large and diverse field. Employment opportunities in each sector exist at manufacturing or research and development companies, both large and small; utilities; government organizations; and nonprofit groups and agencies. Research or education opportunities can be found at universities or trade associations. Because the benefits of renewable energy are a global concern, many employment opportunities can be found outside of the United States.

It is important to note that while employment in the renewable energy industry can be found nationwide, some sectors of the industry tend to be clustered in specific regions of the United States. A good example of this is the wind power industry. Although wind is everywhere, different sections of the United States are windier than other areas. For this reason, wind-related projects tend to be most concentrated in the states of California and Texas, the Pacific Northwest, the Midwest (especially Iowa and Minnesota), and the Mid-Atlantic.

There are a wide variety of employment opportunities in solar energy. Contractors, dealers, distributors, builders, utilities, government agencies, manufacturers, installers, and research and development companies can be found throughout the United States. The Southwest has the greatest potential for solar energy. The Solar Energy Industries Association, a trade organization that supports companies with an interest in solar use, has chapters in 25 states and over 900 business-members located nationwide.

Currently, most geothermal employment opportunities in the United States exist where most geothermal reservoirs are located—in the western states, Alaska, and Hawaii. However, since magma is located everywhere under the earth's surface, better technology and more powerful tools enable geothermal-related projects to be found throughout the country.

Hydropower plants are found throughout the United States. Hydropower projects can be separated into two categories: large hydropower projects run by the federal electric utilities and operated by the Bureau of Reclamation and the Army Corps of Engineers, and nonfederal hydropower dams—about 2,600—licensed by the Federal Energy Regulatory Commission. States that rely heavily on hydropower generation of electricity include Idaho, Washington, Oregon, Maine, South Dakota, California, Montana, and New York.

Biomass is bulky and thus costly to transport. Because of this, bioenergy projects are located where biomass crops are grown. This is a great benefit for many rural areas of the United States since jobs and their economic benefits are kept close to home.

STARTING OUT

Industry associations are a rich source of information, especially when you are looking for your first job. Association Web sites feature the latest industry news, project developments, market forecasts, and government policies. Professional associations, such as the AEE, also offer career advice and job postings on their Web sites.

Many companies recruit on campus or at job fairs. Check with your school's career services center for upcoming fairs in your area. Other good job hunting resources are trade journals, some of which may have job advertisements in their classifieds sections. Check out notable renewable energy publications, such as *Wind Energy Weekly* (http://www.awea.org/wew/index.html) or *Solar Today* (http://www.ases.org/index.php?option=com_content&view=article&id=14&Itemid=22).

Internships are also a great way to get relevant work experience, not to mention valuable contacts. Many of the larger energy companies and nonprofit groups offer internships (either with pay or for course credit) to junior or senior level college students. For example, NREL offers both undergraduate and graduate students the opportunity to participate in its many research and development programs.

ADVANCEMENT

Typical advancement paths depend on the type of position. For example, solar panel installers may advance to positions of higher responsibility such as managing other workers. With experience, they may opt to start their own business specializing in panel installation and maintenance. Engineers may start with a position at a small company with local interests and advance to a position of higher responsibility within that same company (for example, director of research and development). Or they may move on to a larger, more diverse company such as a public utility, whose interests may cover a broader area.

A nontechnical employee with a background in communications, for example, may advance from the human resources department

Did You Know?

- The Romans were among the first to use geothermal energy to heat their homes.
- Albert Einstein was truly a pioneer in renewable energy: In 1921, he won the Nobel Prize in Physics for his groundbreaking experiments in solar power and photovoltaics.
- A solar-powered aircraft set a world record in 1990 when it flew across the United States (in 21 stages) using no fuel at all.
- Hawaii has one of the world's largest wind turbines: It stands 20 stories tall and together its blades are the length of a football field.
- All it takes is a wind speed of about 14 mph to convert wind energy to electricity, and just one wind turbine can produce enough electricity to power up to 300 homes.

Source: "The Natural Solution: Ten Facts About Renewable Energy," by Ana Caistor-Arendar (*The Independent*, May 14, 2007)

of a windmill turbine manufacturing company to handle media and communication requests for a state's energy program. With the proper expertise and credentials, he or she may advance to direct a nonprofit organization representing a sector of the renewable energy industry.

EARNINGS

Very little salary information is available for specific jobs in each subindustry. However, according to the U.S. Department of Labor, the median salary for an electrical engineer was $82,160, as of 2008.

Annual salaries for nontechnical workers vary according to the position, type and size of the employer, and job responsibilities. A typical administrative position would probably pay salaries ranging from $22,000 to $50,000. Those employed by nonprofit organizations tend to earn slightly less than their corporate counterparts. Most employees receive a standard benefits package, including medical insurance, paid vacation and sick days, and a retirement savings program.

WORK ENVIRONMENT

Work environment will vary depending on the industry and the type of position a worker holds. For example, meteorologists in the wind industry may need to travel to distant sites in order to better gauge wind capabilities for a proposed wind turbine project. Solar industry technicians often travel from site to site in order to install or maintain equipment needed for solar projects such as homes, buildings, or thermal generators. Hydropower industry employees may perform much of their work outdoors. Biologists and fisheries managers will work at or near ponds and rivers. Recreation managers may often find themselves developing outdoor walking paths and trails near hydroelectric projects to ensure that vegetation and wildlife are protected. In the geothermal industry, drilling crews work outdoors when they operate heavy drilling tools to locate new reservoirs. Farmers employed by bioenergy companies work outdoors tending their biomass crops. All workers who work outdoors must deal with occasionally extreme weather conditions such as high wind, rain, sleet, snow, and temperature extremes.

Administrative support staff, industry educators, research and development workers, sales and marketing staff, and other nontechnical workers often work indoors in comfortable offices. Many scientists work in laboratories, which are clean, comfortable, and well lit. Most employees work a standard 40-hour week. Important projects or deadlines may require overtime and weekend work.

OUTLOOK

According to a 2008 report by the U.S. Energy Information Administration (EIA), worldwide demand for energy is expected to increase by 44 percent through 2030. EIA predicts that by 2030, oil will supply less of the world's energy needs than it has in the past (32 percent, as opposed to 36 percent today); and that wind and solar power will comprise 11 percent of the global energy supplies. Consumption of oil in the United States is projected to level off and decline by 2030. Presently, about 58 percent of the oil used in the United States is imported. The natural gas share of electricity will stay between 19 and 22 percent through 2030. There is enough coal in the United States to last another 250 years, though that too is limited. Electricity consumption continues to increase faster than conventional methods can produce it, leaving many people and businesses at the mercy of brownouts or blackouts. Political instability in foreign suppliers, increasing costs, and an overreliance on fossil fuels have prompted many to reconsider the potential of renewable energy as a source for unlimited power and fuel.

The solar and wind power industries are the fastest growing sectors of the renewable energy industry. The greatest factor in this growth can be attributed to lower production costs. Better technology and equipment have lowered the cost of wind- and solar-generated electricity. Though European countries continue to show growth in developing renewable energy sources, the United States is also increasing its renewable energy usage. This growth is expected to continue, particularly due to the Obama Administration's focus on clean energy technology. The United States has doubled its wind capacity in the past few years. The American Wind Energy Association estimates that the wind energy industry will triple or even quadruple in the next decade. This is good news for windsmiths, engineers, meteorologists, electricians, and other technical workers.

Solar energy use is already well-established in high-value markets such as remote power, satellites, and communications. Industry experts are working to improve current technology and lower costs to bring solar-generated electricity, hot water systems, and solar-optimized buildings to the public. The manufacturing of PV cell systems also presents many employment opportunities. According to the Energy Information Association (a division of the U.S. Department of Energy), the number of active PV manufacturers and/or importers that ship PV cells and modules increased from 41 companies in 2006 to 46 companies in 2007 to meet international demand for PV systems. Also in 2007, PV cell and module shipments reached a record high, increasing by 53 percent compared to the previous year.

Hydropower is an important renewable energy resource because of its abundance and ability to produce electricity inexpensively without harmful emissions. However, some dams and other water reservoirs have been found to harm fish and wildlife located in or near the project site. The industry has responded to such claims by hiring specialists to protect vegetation and wildlife affected by hydropower projects. Two factors may limit growth in the hydropower industry. First, most potential sites for hydropower projects have already been utilized. Second, the licensing process for hydropower projects is slow and inefficient. License requests must be reviewed and approved by federal and state agencies, which often have a conflict in goals and regulations, making it difficult to obtain a license.

Improved technological advances, such as more powerful drilling tools, have helped the geothermal energy industry grow. Since 1973, geothermal plants have increased capacity by almost 600 percent. Employment opportunities are greatest in the West for the direct use, or drilling, of geothermal energy, and in the Midwest for geothermal heat pumps. However, with advances in technology, employment opportunities will be plentiful throughout the United States. Long delays in obtaining geothermal land leases from the government could hinder the growth of this industry.

Bioenergy is also experiencing steady growth. Interest in bioenergy will not only stem from its electricity potential, but also the biofuels converted from biomass such as ethanol and biodiesel. The U.S. Department of Agriculture estimates that 17,000 jobs are created for every million gallons of ethanol, an important biomass byproduct, produced. Employment opportunities will exist for chemists, engineers, and other agricultural scientists.

Public interest in renewable energy has grown in the last decade. Research has brought better technology, lowered generating costs, and even developed other uses for renewable energy. And the current U.S. government focus on renewable energy products and investments in clean energy solutions is also expected to increase jobs and career opportunities in this field through the next decade.

FOR MORE INFORMATION

Bioenergy

For industry news and updates, general information on bioenergy, contact

> **Renewable Fuels Association**
> One Massachusetts Avenue, NW, Suite 820
> Washington, DC 20001-1401
> Tel: 202-289-3835
> http://www.ethanolrfa.org

General Resources

For information on careers, employment opportunities, certification, membership, and industry surveys, contact

> **Association of Energy Engineers**
> 4025 Pleasantdale Road, Suite 420
> Atlanta, GA 30340-4260
> Tel: 770-447-5083
> http://www.aeecenter.org

For general information on environmental careers, contact
Environmental Career.com
http://environmental-jobs.com

For information on solar tours, energy fairs, industry workshops, or certification, contact
Midwest Renewable Energy Association
7558 Deer Road
Custer, WI 54423-9734
Tel: 715-592-6595
Email: info@the-mrea.org
http://www.the-mrea.org

For more background information on renewable energy, careers, and internships, contact
National Renewable Energy Laboratory
1617 Cole Boulevard
Golden, CO 80401-3305
Tel: 303-275-4099
http://www.nrel.gov

For information on careers, certification and licensing, membership benefits, or local chapters, contact
National Society of Professional Engineers
1420 King Street
Alexandria, VA 22314-2794
Tel: 703-684-2800
http://www.nspe.org

For general information on the renewable energy industry, contact
U.S. Department of Energy Efficiency and Renewable Energy
Mail Stop EE-1
Department of Energy
Washington, DC 20585-0001
Tel: 202-586-9220
http://www.eere.energy.gov

Geothermal
For general information on the geothermal industry and educational teaching guides, contact
Geothermal Education Office
664 Hilary Drive

Tiburon, CA 94920-1446
Tel: 415-435-4574
Email: geo@marin.org
http://www.geothermal.marin.org

For industry news and updates, publications, conferences, career opportunities, and membership information, contact

Geothermal Energy Association
209 Pennsylvania Avenue, SE
Washington, DC 20003-1107
Tel: 202-454-5261
Email: daniela@geo-energy.org
http://www.geo-energy.org

Hydropower

For industry news and updates, publications, conferences, career opportunities, and membership information, contact

National Hydropower Association
25 Massachusetts Avenue, NW, Suite 450
Washington, DC 20001-7405
Tel: 202-682-1700
Email: help@hydro.org
http://www.hydro.org

Solar

For industry news and updates, publications, conferences, career opportunities, and membership information, contact

American Solar Energy Society
2400 Central Avenue, Suite A
Boulder, CO 80301-2862
Tel: 303-443-3130
Email: ases@ases.org
http://www.ases.org

For trade news and updates, publications, conferences, career opportunities, and membership information, contact

Solar Energy Industries Association
575 7th Street NW, Suite 400
Washington, DC 20004-1612
Tel: 202-628-0556
Email: info@seia.org
http://www.seia.org

Wind

For industry news and updates, publications, conferences, career opportunities, and membership information, contact

American Wind Energy Association
1501 M Street, NW, Suite 1000
Washington, DC 20005-1700
Tel: 202-383-2500
Email: windmail@awea.org
http://www.awea.org

Solar Engineers

QUICK FACTS

School Subjects
Computer science
Mathematics
Physics

Personal Skills
Mechanical/manipulative
Technical/scientific

Work Environment
Indoors and outdoors
One or more locations, some travel

Minimum Education Level
Bachelor's degree

Salary Range
$44,484 to $86,209 to $134,969+

Certification or Licensing
Required

Outlook
About as fast as the average

OVERVIEW

Solar engineers work in any number of areas of engineering products that help harness energy from the sun. They may research, design, and develop new products, or they may work in testing, production, or maintenance. They may collect and manage data to help design solar systems. Types of products solar engineers work on may include solar panels, solar-powered technology, communications and navigation systems, heating and cooling systems, and even cars.

HISTORY

People have worshipped the sun and found ways to channel its energy to improve their lives since early times. As far back as 400 B.C., ancient Greeks designed their homes to take advantage of the sun's warmth and light by having the structures face south to capture more heat in the winter. (This is known as "passive solar energy," an old technology that is still used today.) The Romans later improved on these designs by adding more windows to the south side of homes, and by putting glass panes in the windows, which allowed more heat and light into buildings. The Romans were also the first to use glasshouses to grow plants and seeds. And the Greeks and Romans were among the first to use mirrors to reflect the sun's heat to light fires.

Solar cooking is an ancient practice as well, dating at least as far back as the Essenes, an early sect of Jewish people who used the intense desert sun to bake thin grain wafers. In 1767, Swiss Naturalist Horace-Bénédict de Saussure created the first solar oven—an insulated, glazed box with a glass-paned cover, which reached temperatures of 190-degrees Fahrenheit. In the 1950s, to aid communities located near deserts, the United Nations and other agencies funded studies of solar cooking to determine if it was a viable way to reduce reliance on plant life for fuel. The studies proved solar cooking was feasible, and so the UN provided further funding for programs to introduce wooden solar cookers to communities in need, such as in locations where firewood was scarce. Despite the benefits of the cookers, however, most groups ended up sticking with their old cooking methods and turned the cookers into firewood.

Solar cooking is back in force today, though. Solar ovens can now reach temperatures as high as 400 degrees Fahrenheit. Many hobbyists, inventors, and designers have fine-tuned the designs of solar ovens over the years, some turning them into marketable products. And the UN's solar cooking idea has been resurrected. In 2006, the nonprofit organizations Jewish Watch International, KoZon Foundation, and Solar Cookers International successfully launched a program to bring solar cookers to Darfur refugees. Civil war started in 2003 in Darfur (located in Western Sudan, Africa) and violence has raged in the years since. As of 2008, at least 200,000 people had lost their lives and 2.5 million had been displaced. As simple an idea as it seems, solar cookers could actually save lives, because women and girls would no longer need to leave the safety of numbers to head off alone in search of firewood.

Other important developments and highlights in solar energy include:

- 1839: French Physicist Alexandre-Edmond Becquerel discovers the photovoltaic effect (from Greek, "photo" meaning light, and "voltaic" meaning electricity), which is when certain materials produce small amounts of electricity when exposed to light.

- 1883: American Inventor Charles Fritts creates the first working solar cell (a device that converts solar energy directly into electricity), by coating semiconductor-material selenium with a thin layer of gold. (Efficiency rate was a miniscule 1 to 2 percent.)
- 1921: Albert Einstein wins the Nobel Prize in Physics for his theories about the photovoltaic effect.
- 1947: Because energy is scarce during World War II, the United States builds more passive solar buildings. The book *Your Solar House*, published by Libby-Owens-Ford Glass Company, also sparks interest in solar energy.
- 1954: Bell Laboratories invents the "solar battery"—the first practical silicon solar cell with a sunlight energy-conversion efficiency of about 6 percent. Bell demonstrates its invention by powering a toy Ferris wheel and a solar-powered radio transmitter.
- 1955:
  - Mechanical Engineer Frank Bridgers designs the first commercial office building to use solar water heating and passive solar design. The Solar Building, located in Albuquerque, is now a designated historic landmark.
  - Western Electric begins selling commercial licenses for silicon photovoltaic technologies.
- 1959: Explorer VI satellite is launched with a photovoltaic (PV) array of 9,600 solar cells. ("Array" means an interconnected system of PV modules that function as one unit that produces electricity.)
- 1960: Hoffman Electronics achieves PV cells with 14 percent efficiency.
- 1963: Sharp starts producing PV modules. (A module is a group of individual PV cells that are used to harness solar radiation for energy.)
- 1969: A solar furnace is built in Odeillo, France, featuring an eight-story parabolic mirror to capture sun energy.
- 1970s:
  - The Solar Energy Industries Association and the Solar Energy Research Institute (now the National Renewable Energy Laboratory) are formed.
  - Thin-film photovoltaic research begins. (In thin-film solar technology, non-silicon semiconductor materials such as copper, indium, gallium and selenium (CIGS) are used to create photovoltaic cells that convert sunlight into electricity.)

- Solar cell prices drop from about $100 per watt to about $20 per watt.
- 1980s:
  - ARCO Solar becomes the first company to produce more than 1 megawatt (1000 kilowatts) of photovoltaic modules in one year.
  - American Inventor Paul MacCready builds the Solar Challenger, the first solar-powered aircraft, and flies it from France to England across the English Channel.
  - Solar One, a 10-megawatt solar-power demonstration project, begins operations (1982–88), and proves the feasibility of power tower systems.
- 1990s:
  - National Renewable Energy Laboratory produces a solar cell with 30 percent conversion efficiency.
  - The tallest skyscraper built in the 1990s in New York City is completed. The building–4 Times Square–has more energy-efficient features than any other commercial skyscraper, including building-integrated photovoltaic panels on floors 37 through 43 (on the south- and west-facing facades) to produce part of the building's power.
  - Worldwide-installed PV capacity reaches 1000 megawatts.
- 2000s:
  - Home Depot begins selling residential solar power systems in San Diego, California.
  - Helios, NASA's solar-powered aircraft, sets a new world altitude record for non-rocket-powered craft: 96,863 feet (more than 18 miles up).
  - University of Colorado students build an energy-efficient solar home for the Solar Decathlon, a competition sponsored by the U.S. Department of Energy, and win first prize. (Students integrated aesthetics and modern conveniences with maximum energy production and optimal efficiency.) The houses are transported to the National Mall in Washington, D.C.

THE JOB

Solar engineering, while an ancient practice, is still a relatively new industry that has caught more mainstream attention only within the past 20 years. With forecasts of fossil fuels' eventual extinction and

Solar engineers install solar panels on the roof of a building. The panels provide power for the building by converting light into electricity with no waste and no harmful emissions. *Paul Rapson/Photo Researchers, Inc.*

the focus shifting to sustainable business practices, more engineers are researching and developing solar-powered products as a means to conserve energy.

There are two types of solar energy: *passive solar energy* and *active solar energy*. Passive solar energy, as the name suggests, means that no mechanical devices are needed to gather energy from the sun. Positioning buildings to face the sun is one example of passive solar energy. In direct contrast, mechanical devices are used for active solar energy—to collect, store, and distribute solar energy throughout buildings. For instance, mechanical equipment such as pumps, fans, and blowers are used to gather and distribute solar energy to heat the space inside a home. Active solar energy is just one area in which solar engineers work. They help create active solar-space heating systems that are liquid (e.g., water tanks) or air based (e.g., rock bins that store heat), and active solar-water heating systems that use pumps to circulate and heat fluids.

Solar engineers are frequently electrical, mechanical, civil, chemical, or even petroleum engineers who are working on solar projects and designing photovoltaic systems. Solar engineer John Gardner says, "Solar engineers could work on everything from designing materials for new solar cells, designing mounting systems, collecting and managing environmental data used to design solar systems, designing actual solar installations to researching and developing future systems and components." Gardner works for Solar Renewable Energy, a company that provides energy-saving services and products—including solar—to customers.

Solar engineers may be responsible for such things as reviewing and assessing solar construction documentation; tracking and monitoring project documentation; evaluating construction issues; meeting with other engineers, developers, and investors to present and review project plans and specifications; participating in industry forums; and possibly even dealing with clients directly. One general requirement for most solar engineering positions is a working knowledge of mechanical and electrical engineering, and an understanding of a range of engineering concepts (such as site assessment, analysis, and design, and energy optimization).

REQUIREMENTS
High School
Take classes in math (e.g., algebra, calculus, geometry), science, natural science, communications, and computers. Engineering schools tend to favor students who have taken advanced placement and honors classes, so do your best to pursue coursework at this high level.

INTERVIEW

John Gardner, Solar Engineer

Q. What do you do in your job?

A. My day-to-day tasks include the preliminary design of solar PV systems, developing estimates for system cost, preliminary bills of material, and stamp (PE, Professional Engineer) certified design drawings.

Q. How long have you been a solar engineer?

A. Professionally, I have been doing this for a little over year. I have been a professional electrical engineer for 37 years.

Q. What is your work background? And what sparked your interest in solar engineering?

A. After college, I worked in industrial manufacturing for the petrochemical industry. At one point, our company worked on heat exchangers for flat-plate solar thermal panels. I have researched solar PV and wind systems for almost 10 years and installed a system on my house five years ago—consisting of a 1-kilowatt wind generator, a 2-kilowatt PV system, and a solar hot-water heater.

Q. What do you like most about your work? And what do you like least?

A. The thing I like most (which seems a little like a fairy tale) is that I get to come to work every day and actually design solar systems that people are going to install on their homes or businesses. The least fun part is working within short deadlines, which are typical of any business.

Q. Is there anything that surprised you, or that you didn't expect, about this work or field?

A. What has surprised me the most is the large number of people really interested in solar. Not everyone can afford to have a system installed on his or her home now, but they are ready to do so as soon as they can.

Postsecondary Training

There are about 1,830 ABET-accredited (Accreditation Board for Engineering and Technology) colleges and universities that offer bachelor's degrees in engineering, according to the U.S. Department of Labor. Most solar engineers have a bachelor's of science in an engineering specialty, e.g., electrical, civil, mechanical, or chemical engineering. Engineering programs typically include mathematics, physical and life sciences, and computer or laboratory courses. Classes in social sciences or humanities are usually required as well. Many companies prefer to hire engineers with master's of science degrees, so those who pursue advanced degrees may have better odds of securing work.

Certification or Licensing

All 50 states and the District of Columbia require engineers who offer their services to the public to be licensed as professional engineers (PEs). To be designated as a PE, engineers must have a degree from an engineering program accredited by the Accreditation Board for Engineering and Technology, four years of relevant work experience, and successfully complete the state examination.

Other Requirements

A passion for solving problems is a key characteristic of all engineers, and particularly of those who work on renewable energy projects. Solar engineers team up with a wide variety of people—from management, fellow engineers, designers, and construction professionals, to developers, clients, investors, and more—so it's essential to have strong communication skills, a flexible attitude, and the ability to get along well with others.

EXPLORING

Learn more about solar energy by reading magazines such as *Home Power* and *Solar Today*, and visit Web sites like Build It Solar (http://www.builditsolar.com) to find all sorts of links to solar projects, designs, and experiments that you might even be interested in doing yourself. You can set up a small solar system at home and see firsthand how it works. To get an idea about the types of engineering jobs that are out there, visit such Web sites as Intech.net (http://www.intech.net) and Simply Hired (http://www.simplyhired.com).

EMPLOYERS

Solar engineering is a growing field. While many engineers are working on solar projects, there are no statistics available yet regarding the number of solar engineers who are working full time in America. According to the U.S. Department of Labor, in 2006, there were 256,000 civil engineers, 227,000 mechanical engineers, 153,000 electrical engineers, 30,000 chemical engineers, and 17,000 petroleum engineers employed in the United States.

"A wide variety of companies would hire a 'solar engineer,'" John Gardner says. "An installation company would require an engineer to design jobs and stamp design drawings. A manufacturing company would hire an engineer to develop manufacturing and production systems to reduce costs and improve quality in the manufacture of system components." Power systems companies, solar cell and module manufacturers, solar panel companies, and companies that provide energy-saving services (such as heating and cooling systems, energy audits, etc.) to commercial and residential customers are just a few examples of the types of companies that hire engineers to work on solar projects.

The Sunny Side of Big Business

Rising fossil fuel prices and tighter regulations on carbon emissions are pushing companies to explore using renewable energy. Some companies are setting up solar-thermal plants, which use mirrors to heat fluids that generate steam. The steam then drives turbines that can provide electricity to multiple users, all at standard prices. Frito-Lay, for example, is exploring the use of alternative energy such as solar power to produce its snacks. With the help of the National Renewable Energy Laboratory (NREL), the company chose its Casa Grande, Arizona, plant, which was built in 1984, as the test site for manufacturing its products with alternative energy. The desert location won out over others because of the abundant sunshine and Arizona's focus on water conservation. Frito-Lay is retrofitting the company with equipment to recycle waste and generate power, by adding such things as high-tech filters to clean the water used to rinse potatoes so it can be reused, and 50 acres of solar concentrators to generate solar power to cook its chips. The renovation should be completed by 2010, and the company plans to use Casa Grande as a model for other plants across the country.

Most engineers, in general, work in architectural, engineering, and related services. Some work for business consulting firms, and manufacturing companies that produce electrical and electronic equipment, business machines, computers and data processing companies, and telecommunications parts. Others work for companies that make automotive electronics, scientific equipment, and aircraft parts; consulting firms; public utilities; and government agencies. Some may also work as private consultants.

STARTING OUT

See if you can get an internship with a company that provides solar energy services. You can also learn more about the industry by visiting the Web sites of professional associations such as the Institute of Electrical and Electronic Engineers (IEEE), American Wind Energy Association, and the American Solar Energy Society. If there's an upcoming meeting or event in your area, it may be a good opportunity to meet solar energy professionals, find out about the latest trends, and learn where the job market is heading.

ADVANCEMENT

Solar engineers who work for companies can advance by taking on more projects, managing more people, and moving up to senior-level positions. They may start their own companies and expand their business by offering more services and opening up branches in other locations. They may also teach at universities and write for various publications.

EARNINGS

Solar power engineers averaged about $86,209 per year in 2009, according to Simply Hired. Salaries can vary depending on location. Solar engineers in North Dakota, for example, earned about $71,000 per year, whereas solar engineers in New York City had annual incomes of about $107,000.

Salaries for solar engineers can also vary depending on the type of engineering work they do. According to Salary.com, in 2009, entry-level electrical engineers earned annual salaries ranging from $48,893 to $67,789, while those with more years of experience brought home between $99,191 to $134,969 or more per year. Beginning civil engineers earned slightly lower salaries: $44,484 to $62,526. And mechanical engineers with some work experience had annual earnings ranging from $58,355 to $85,022 or higher.

WORK ENVIRONMENT

Solar engineers may work indoors or outdoors, depending on the project. Work hours are generally 40 per week, with longer hours required when projects near deadline dates. Solar engineers may work in office buildings, laboratories, or industrial plants. They may spend time outdoors at solar power plants, and may also spend time traveling to different plants and worksites in the United States as well as overseas.

OUTLOOK

The U.S. Department of Labor forecasts average growth in employment of engineers through 2016, and slow employment growth for electrical engineers specifically. But engineers working in renewable energy can look forward to better odds of finding work in the years to come, especially as more governments invest money into alternative-energy research and development.

The solar engineering field will get "bigger" and "more specialized," according to John Gardner. "Most of the current systems are smaller scale installations," he says. "But there are many new utility scale systems being installed. This will require knowledge of utility interconnection at higher voltages than typically seen in residential and commercial."

According to Engineering.com, "The solar power market has grown significantly in the past decade," and will continue to do so. Solarbuzz (http://www.solarbuzz.com) also forecasts growth in the solar industry over the next few years, predicting industry revenue as high as $31.5 billion by 2011—nearly three times that of the $10.6 billion solar-power revenue in 2006.

FOR MORE INFORMATION

Find solar-power industry news, career information, and listings for events and conferences by visiting the Web sites of the following associations:

American Solar Energy Society
2400 Central Avenue, Suite A
Boulder, CO 80301-2843
Tel: 303-443-3130
Email: ases@ases.org
http://www.ases.org

Solar Electric Power Association
1220 19th Street, NW, Suite 401
Washington, DC 20036-2405

Tel: 202-857-0898
Email: info@solarelectricpower.org
http://www.solarelectricpower.org

Solar Energy Industries Association
575 7th Street, NW, Suite 400
Washington, DC 20004-1612
Tel: 202-682-0556
http://www.seia.org

For information on engineer careers and educational programs, contact the following groups:

Institute of Electrical and Electronics Engineers
2001 L Street, NW, Suite 700
Washington, DC 20036-4910
Tel: 202-785-0017
Email: ieeeusa@ieee.org
http://www.ieee.org

National Renewable Energy Laboratory
1617 Cole Boulevard
Golden, CO 80401-3305
(303) 275-3000
http://www.nrel.gov

Visit Solarbuzz's Web site for industry news and job postings.

Solarbuzz
PO Box 475815
San Francisco, CA 94147-5815
Tel: 415-928-9743
Email: info@solarbuzz.com
http://www.solarbuzz.com

Find out about workshops in sustainable living by visiting

Solar Living Institute
PO Box 836
13771 South Highway 101
Hopland, CA 95449-9607
Tel: 707-472-2450
http://www.solarliving.org

Wind Power Engineers

 **QUICK FACTS**

School Subjects
Mathematics
Physics

Personal Skills
Analytical/creative
Technical/scientific

Work Environment
Indoors and outdoors
One or more locations, some travel

Minimum Education Level
Bachelor's degree

Salary Range
$44,484 to $67,789 to $134,969+

Certification or Licensing
Required

Outlook
About as fast as the average

OVERVIEW

Wind power engineers work in the areas of design, construction, operation, and maintenance of wind power machinery. Depending on their engineering discipline, they may research and test locations and equipment, design wind turbines and wind farms, oversee construction, and test the functioning of machinery.

HISTORY

Windmills were first used as far back as 200 B.C. in China to pump water, and in Persia and the Middle East to grind grain. Starting in the mid-16th century, windmills were used widely in Holland, which is below sea level, to drain land and keep it dry. They were also used in the timber and paper industries. By 1890, industrialization caused the development of larger windmills, known as wind turbines, in Denmark. And by the early 1900s, windmills served a dual purpose in the United States: to pump water for farms and ranches, and provide electricity to homes and businesses. The largest, electricity-producing wind turbine in the United States was installed on a summit known as Grandpa's Knob in Castleton, Vermont, providing electricity during World War II to about 1,000 homes. The development of steam engines and electric motors gradually replaced the need for traditional windmills, however, and the industry started declining in the 1920s.

The fuel oil crisis in the 1970s renewed interest in wind power as an alternative energy source, and by the 1980s, California was leading the way in the United States in wind energy development. Wind energy research and development has continued throughout the states in the years since. According to the American Wind Energy Association, as of 2007, Texas had taken over the lead in wind power development, with California coming in at a close second, and Iowa, Minnesota, and Washington State trailing behind. The downfalls to this technology are few: wind turbines are prone to damage by lightning, damage by (and to) birds flying into the blades, and people who are nearby have complained about the noise from the turning of the blades. The advantages of this clean, renewable energy source outweigh these: no greenhouse gas emissions or other pollutants; each wind turbine (while tall) only takes up a small plot of land, so the land beneath it can still be used; remote areas that cannot connect to electrical power grids can use wind turbines for electricity.

THE JOB

Land, water, weather, vegetative cover, and other factors influence wind strength and frequency. Wind energy is actually a form of solar energy. The sun heats different parts of the earth at different rates. When hot air rises, causing a drop in the atmospheric pressure at the earth's surface, cool air comes in, and that's when wind is created. The motion of the wind has energy in it, and wind turbines harvest this energy and convert it first into mechanical energy, and then into electricity. Wind electric turbines have two basic designs: vertical-axis ("egg-beater") style, or horizontal-axis (propeller) style. Wind turbines come in different sizes: small turbines are used for residences and small businesses. Horizontal-axis machines dominate the utility-scale (meaning they have 100-kilowatt capacity or larger) turbines in the global market. Wind turbines are tubular, steel towers and the blades are usually made of fiberglass-reinforced polyester or wood-epoxy. A wind power plant or wind farm consists of a group of wind turbines set far apart from each other on a large, windy tract of land. Northern California has the largest concentration of wind turbines in the United States, including the Altamont Pass Wind Farm in Central California: Built in the 1970s to combat high fuel costs, it consists of 4,900 small wind turbines and is one of this country's earliest wind farms.

A wind power engineer stands in front of wind turbines that power his home.
AP Photo/The Hutchinson News, Patrick Traylor

Wind power engineers bring different engineering disciplines to wind power projects. According to Peder Hansen, executive vice president of Northstar Wind Towers, LLC, who has a mechanical engineering background and more than 25 years' experience in the wind energy field, many different engineers work in the wind power business:

- *Aerodynamics engineers* work on blade designs and airflow studies for more efficient ways to harness the power of the wind.
- *Electrical engineers* ensure that the energy from the wind is converted in the best possible manner into electrical energy through generators and transformers of various sorts. They are also responsible for getting the power from the turbine to the grid.
- *Electronics engineers* design the controllers and converters that control the functions of the machine.
- *Mechanical engineers* work on gearboxes, bed-plates, bearings, towers, bolts, etc. (This is a large area of expertise and requires several specialists.)

- *Civil engineers* design foundations for the turbines and roads that connect the turbines to each other. These engineers are also responsible for overhead or underground cable work.
- *Composite engineers* work closely with the aerodynamics engineers to ensure that the turbine blades (which are made of fiberglass) can in fact be made, and can be made efficiently.

Wind engineers may also work in the areas of sales, making presentations and liaising with current and prospective customers. Peder Hansen says, "Technical sales is my forte, ensuring that the customers get the right product for the project location." His work involves evaluating wind data and gathering information from customers, and then helping customers make the right choices regarding turbine specifications and tower heights. A big part of the job also entails making sure he has all of the latest market information to enable the company to make the right strategic decisions. In the past, one of the most challenging, and interesting, projects he's worked on was when he was part of a multidiscipline engineering team that acquired the rights to a 1.5-megawatt wind turbine in Germany. They had to convert it to a machine that was capable of operating in North America, which entailed "configuring new controls, algorithms, gearbox ratios, generator speeds, frequency ranges, as well as reconfiguring the bed-plate to withstand much more extreme temperature swings than originally intended. It also included finding domestic vendors for all of these parts and testing them in lab and real-life settings."

Bjorn S. Gullaksen, PMP (certified project management professional), managing senior associate at Stantec, a Portland, Oregon-based organization that provides sustainable design and consulting services to a global clientele, describes the steps involved in wind engineering as follows:

Multiple wind turbines, called Wind Turbine Generators (WTGs), are installed on wind farms, requiring several acres of land and multiple permits before they can be put into service providing electricity. Several wind-engineering disciplines are required before this can take place:

1. First, an engineer looks at the site where the wind farm will be constructed to evaluate the wind's strength and frequency.

2. Next, an *environmental engineer* needs to look at the site to see if there are any endangered species present and to make sure that constructing the wind farm will not harm the environment.
3. Once the green-light is received to build the wind farm, the design needs to be done, which involves electrical, civil, and structural engineers designing the power grids, roads, structures, etc. Also, *mechanical engineers* are used to design the WTGs themselves (e.g., gearboxes, breaks, etc.).
4. When construction starts, a *construction engineer* is needed to oversee the construction and handle contractors' questions.
5. Before the wind farm can be energized, an *electrical engineer* and a *testing and commissioning engineer* test all of the equipment, ensuring that all parts function the way they should.

Bjorn Gullaksen is an electrical engineer who learned about how electricity is produced and delivered to homes and industries from his previous work with a large electric utility. His current work involves running a consulting business that provides wind farm design services. As managing senior associate at Stantec, his responsibilities include overseeing the work of engineers and project managers, and making sure customers receive quality service. He says the most challenging projects have been those that are difficult to design, such as a wind farm located up in the mountains (like those in California), where access is challenging. Gullaksen has been in the wind consulting business for two years; he was inspired to get into it by the need to combat global warming by producing electricity via renewable energy. "Wind is the friendliest source of energy to achieve this goal," he says. "The most satisfaction I get out of my job is the fact that I am helping to provide the world with a better environment for us to survive in."

REQUIREMENTS
High School
Engineers need strong math and science skills, as well as solid verbal and written communication abilities. Course work in algebra, trigonometry, calculus, biology, physics, chemistry, computers, English, and business will provide a solid foundation for college.

Postsecondary Training

Wind power engineers have different specialties and therefore pursue different undergraduate degrees, depending on their areas of interest. Most hold bachelor's degrees in any one of these engineering disciplines: aerodynamics, environmental, electrical, electronics, civil, mechanical, or structural. There are about 1,830 ABET-accredited (Accreditation Board for Engineering and Technology) colleges and universities that offer bachelor's degrees in engineering, according to the U.S. Department of Labor. Many schools are starting to offer programs in renewable energy and course work in energy systems engineering. Engineers who pursue advanced degrees increase their hiring potential.

Certification or Licensing

All 50 states and the District of Columbia require engineers who offer their services to the public to be licensed as professional engineers (PEs). To be designated as a PE, engineers must have a degree from an engineering program accredited by the Accreditation Board for Engineering and Technology, four years of relevant work experience, and successfully complete the state examination. Engineers who wish to advance to more senior positions can further their careers by securing certification in business management from professional associations.

Other Requirements

The ability to gather and analyze information is essential in this work, as are people skills. Engineers frequently work with teams, so strong communication skills and a flexible attitude are required to accomplish goals. Engineers are also always learning; education does not stop once they have their degree in hand. In engineering, education is a career-long process. Wind power engineers keep up with alternative energy trends and policies by reading books and magazines, taking classes, and attending conferences.

EXPLORING

Visit the Web sites of the American Wind Energy Association (http://www.awea.org) and Wind Power (http://www.windpower.com) to learn more about the field. Keep up with industry news by reading magazines such as *Windpower Monthly, North American Wind Power, Renewable Energy World,* and *North American Clean Energy.* Peder Hansen, who developed a "lifelong passion for wind turbines and alternative sources of energy" when he saw the first experimental wind turbine go up in his hometown in Denmark, recommends reading Lester R. Brown's book *Plan B*, and Van Jones's book *The Green Collar Economy*.

EMPLOYERS

There were 256,000 civil engineers, 227,000 mechanical engineers, 153,000 electrical engineers, and 30,000 chemical engineers employed in the United States in 2006, according to the U.S. Department of Labor. Wind engineers work predominantly for wind turbine companies. "Many developers also have their own engineering staff to assist them in due diligence when choosing a turbine for a particular installation," Peder Hansen says. "There are a few large engineering and certification entities who also hire wind engineers, but these tend to be on a more theoretical level." Hansen has worked for such companies as GE Energy, Vestas, and Valmont. Wind engineers may also work for electrical utilities, manufacturers, consultancies, or even law firms.

STARTING OUT

An apprenticeship is an excellent way to learn the job firsthand. Peder Hansen started as a shop-floor assembly worker and then moved into installing and troubleshooting wind power machines. Attend trade shows such as the annual American Wind Energy Association's Windpower show, where you'll have the opportunity to meet professionals in the field and learn firsthand what's new in wind energy and renewable energy. And Bjorn Gullaksen says, "Keep studying calculus, trigonometry, and differential equations, as well as work on becoming an efficient writer." Honing these skills will help advance you on the path to engineering. You can also learn more about the types of wind energy projects companies are hiring for by visiting employment Web sites such as Wind Industry Jobs (http://www.windindustryjobs.com) and Jobs in Wind Power (http://www.jobsinwindpower.com).

ADVANCEMENT

Junior-level engineers can advance by taking on more complicated projects and moving up to senior management positions. They may be responsible for overseeing more employees and managing more facilities within the region or within different regions. Engineers with certification in business and project management advance to positions of greater responsibility and authority. Those with years of experience in the field start their own consulting firms and branch out into other services. Some may lecture, teach at universities, and write books and articles.

EARNINGS

Salaries for wind power engineers vary depending upon their specialty and the project. According to Salary.com, in 2009, entry-level electrical engineers earned annual salaries ranging from $48,893 to $67,789, while those with more years of experience brought home between $99,191 to $134,969 or more per year. Beginning civil engineers earned slightly lower salaries: $44,484 to $62,526. And mechanical engineers with some work experience had annual earnings ranging from $58,355 to $85,022 or higher. Aerospace engineers had salaries ranging from $58,130 to $134,570.

WORK ENVIRONMENT

Wind power engineers work indoors in offices and outdoors at wind power farms. They may 40 hours per week or more, depending on project deadlines. During certain phases of projects, some engineers will be on call 24/7, and will need to be accessible and available at any time of day or night. They travel for meetings and conferences and may need to relocate for long-term projects or permanent work.

OUTLOOK

States across America are upping their mandates for renewable energy use. A 2009 report by the National Research Council projected that the United States could get up to 20 percent of its power from non-hydro renewable energy by 2020. With the focus turning more toward clean energy, the demand for engineers in this field is expected to rise. The U.S. Department of Labor forecasts average growth in employment of engineers through 2016. Engineers specializing in renewable energy such as wind and solar power, however, should have more employment opportunities within the next decade. Governments and private industries will continue to invest more money in wind power research and development in the coming years. One example of this is the construction of Cape Wind, the first offshore wind farm in the United States, located off the coast of Cape Cod, Massachusetts, which is expected to be completed by 2010.

FOR MORE INFORMATION

Visit the Web sites of the following associations for industry news, upcoming events and conferences, and membership information:

American Association for Wind Energy
c/o Cermak Peterka Petersen, Inc.
1415 Blue Spruce Drive, #3
Fort Collins, CO 80525-2003
Tel: 970-498-2334
Email: aawe@aawe.org
http://www.aawe.org

American Society of Civil Engineers
1801 Alexander Bell Drive
Reston, VA 20191-4400
Tel: 800-548-2723
http://www.asce.org

For information on engineer careers and educational programs, contact these groups:

Institute of Electrical and Electronics Engineers
2001 L Street, NW, Suite 700
Washington, DC 20036-4910
Tel: 202-785-0017
Email: ieeeusa@ieee.org
http://www.ieee.org

National Renewable Energy Laboratory
1617 Cole Boulevard
Golden, CO 80401-3305
(303) 275-3000
http://www.nrel.gov

Further Reading

Ashby, Darren. *Electrical Engineering 101: Everything You Should Have Learned in School...but Probably Didn't*. 2d ed. Burlington, Mass.: Newnes, 2008.

Bethscheider-Keiser, Ulrich. *Future Cars: Green Designed: Bio Fuel, Hybrid, Electrical, Hydrogen, Fuel Economy in All Sizes and Shapes*. Ludwigsburg, Germany: Avedition, 2008.

Boschert, Sherry. *Plug-in Hybrids: The Cars That will Recharge America*. Gabriola Island, Canada: New Society Publishers, 2006.

Boyle, Godfrey. *Renewable Energy*. New York: Oxford University Press, 2004.

da Rosa, Aldo V. *Fundamentals of Renewable Energy Processes*. 2d ed. Burlington, Mass.: Academic Press, 2009.

Das, Braja M. *Principles of Geotechnical Engineering*. CL-Engineering, 2009.

Fenton, Gordon A. and D.V. Griffiths. *Risk Assessment in Geotechnical Engineering*. Hoboken, N.J.: Wiley, 2008.

Fell, Robin, et al. *Geotechnical Engineering of Dams*. London: Taylor & Francis, 2005.

Gevorkian, Peter. *Alternative Energy Systems in Building Design*. New York: McGraw-Hill Professional, 2009.

Gibilisco, Stan. *Alternative Energy Demystified*. New York: McGraw-Hill Professional, 2006.

Gipe, Paul. *Wind Power: Renewable Energy for Home, Farm, and Business*. Rev. ed. White River Junction, Vt.: Chelsea Green Publishing Company, 2004.

Goswami, D. Yogi. *Principles of Solar Engineering*. New York: CRC Press, 2000.

Hambley, Allan R. *Electrical Engineering: Principles and Applications*. 4th ed. Saddle River, N.J.: Prentice Hall, 2007.

Higman, Christopher and Maarten van der Burgt. *Gasification*. 2d ed. Burlington, Mass.: Gulf Professional Publishing, 2008.

Hyne, Norman J. *Nontechnical Guide to Petroleum Geology, Exploration, Drilling and Production*. 2d ed. Tulsa, Okla.: PennWell Books, 2001.

Leitman, Seth. *Build Your Own Plug-In Hybrid Electric Vehicle*. New York: McGraw-Hill/TAB Electronics, 2009.

Lyons, William C. and Gary J. Plisga. *Standard Handbook of Petroleum and Natural Gas Engineering*. 2d ed. Burlington, Mass.: Gulf Professional Publishing, 2004.

Manwell, James. F., et al. *Wind Energy Explained*. Hoboken, N.J.: Wiley, 2002.

Masters, Gilbert M. *Renewable and Efficient Electric Power Systems*. Hoboken, N.J.: Wiley-IEEE Press, 2004.

Miller, Bruce G. *Coal Energy Systems (Sustainable World)*. Burlington, Mass.: Academic Press, 2004.

Murray, Raymond L. *Nuclear Energy: An Introduction to the Concepts, Systems, and Applications of Nuclear Processes*. 6th ed. Chicago: Butterworth-Heinemann, 2008.

Nerad, Jack R. *The Complete Idiot's Guide to Hybrid and Alternative Fuel Vehicles*. New York: Alpha, 2007.

Olah, George A., et al. *Beyond Oil and Gas: The Methanol Economy*. Hoboken, N.J.: Wiley-VCH, 2006.

Rajapakse, Ruwan. *Geotechnical Engineering Calculations and Rules of Thumb*. Chicago: Butterworth-Heinemann, 2008.

Rizzoni, Giorgio. *Fundamentals of Electrical Engineering*. New York: McGraw-Hill Science/Engineering/Math, 2008.

Shultis, J. Kenneth and Richard E. Faw. *Fundamentals of Nuclear Science and Engineering*. 2d ed. New York: CRC Press, 2007

Sims, Ralph. *The Brilliance of Bioenergy: In Business and Practice*. London: Earthscan Publications Ltd., 2002.

Soetaert, Wim and Erik Vandamme, eds. *Biofuels*. Hoboken, N.J.: Wiley, 2009.

Tiwari, G.N. *Solar Energy: Fundamentals, Design, Modeling and Applications*. New Delhi, India: Narosa, 2002.

Vaitheeswaran, Vijay and Iain Carson. *ZOOM: The Global Race to Fuel the Car of the Future*. New York: Twelve, 2008.

Wall, Judy D., Caroline S. Harwood and Arnold Demain, eds. *Bioenergy*. Washington, D.C.: ASM Press, 2008.

Williams, A. *Combustion and Gasification of Coal*. New York: CRC Press, 2000.

Index

Entries and page numbers in **bold** indicate major treatment of a topic.

A

ABET. *See* Accreditation Board for Engineering and Technology
accountants 135. *See also* renewable energy workers
Accreditation Board for Engineering and Technology (ABET) 18, **28–29**, 50, 71, 80, 102, 155, 165
"active solar energy" 153
advertising workers 135. *See also* renewable energy workers
AEE. *See* Association of Energy Engineers
aerodynamics engineers 162. *See also* wind power engineers
agricultural scientists 135. *See also* renewable energy workers
agricultural technicians 4. *See also* bioenergy/biofuels workers
AISES. *See* American Indian Science and Engineering Society
Alaska 118, 139
Albuquerque, New Mexico 150
Alcoa Aluminum 67
Altamont Pass Wind Farm (California) 161
American Academy of Environmental Engineers 18, 50, 71
American Association for Wind Power 168
American Association of Petroleum Geologists 107, 120
American Chemical Society 11, 21
American Coal Foundation 19, 21
American Council for an Energy-Efficient Economy 12, 63
American Indian Science and Engineering Society (AISES) 103, **107–108**
American Institute of Biological Sciences 11
American Institute of Chemical Engineers 21
American Institute of Constructors 11
American Nuclear Society (ANS) 82, 83, 85, 97
American Petroleum Institute 44, 101, 108, 111, **120–121**
American Public Power Association 129
American Recovery and Reinvestment Act 131
American Society for Engineering Education (ASEE) 12
American Society for Nondestructive Testing 97
American Society of Agronomy 11
American Society of Civil Engineers (ASCE) 52, 168
American Solar Energy Society (ASES) vii, 146, 157, 158
American Telephone and Telegraph (AT&T) 23
American Wind Energy Association (AWEA) 132, 143, 147, 157, 161, 165, 166
ANS. *See* American Nuclear Society
Appleton, Wisconsin 66
appraisal well **112–113**
Arab/Israeli War 110
architects 131, 134. *See also* renewable energy workers
ARCO Solar 151
Arctic Circle 104
ASCE. *See* American Society of Civil Engineers
ASEE. *See* American Society for Engineering Education
ASES. *See* American Solar Energy Society
Association for Operations Management 12

Association of Energy Engineers (AEE) 39, 44, 74, 137, 140, 144
AT&T. *See* American Telephone and Telegraph
Austin-Healey Sprite 55
auxiliary equipment operators 124. *See also* power plant workers
AWEA. *See* American Wind Energy Association

B

Beck Company 55
Becquerel, Alexandre-Edmond 131, 149
Bell, Alexander Graham 22, 23
Bellingam, Washington 67
Bell Laboratories 150
Bell Telephone Company 23
biochemists 135. *See also* renewable energy workers
bioenergy 1, 135
bioenergy/biofuels workers vii, **1–12**
bioenergy plant scientists 4. *See also* bioenergy/biofuels workers
biofuels 1
Biofuels Digest 7
biofuels plant managers 5. *See also* bioenergy/biofuels workers
biofuels product managers 4. *See also* bioenergy/biofuels workers
biological technicians 4. *See also* bioenergy/biofuels workers
biologists 4, 134, 135. *See also* bioenergy/biofuels workers; electricians; renewable energy workers
biomass 1, 140
Blatch, Nora Stanton 47
Bonneville Lock and Dam 70, 72
Bonneville Power Administration 67
The Bourne Identity 55
BP Company 109
Braille Battery Company 56
Bridgers, Frank 150
Bristol Resources 119
Brown, Lester R. 166
Buffalo, New York 66
Build It Solar (Web site) 155
businesspeople 131. *See also* renewable energy workers

C

California 62, 102, 104, 118, 139, 161
California Geoprofessionals Association 52–53
Cape Wind (Cape Cod, Massachusetts) 167–168
Car and Driver 60
carpooling 2
CEM. *See* certified engineering manager
central control room operators 124. *See also* power plant workers
certification. *See specific careers*
certified engineering manager (CEM) 137
certified project manager professional (PMP) 163
chemical engineers 134. *See also* renewable energy workers
chemical technicians 114. *See also* petroleum technicians
chemists 135. *See also* renewable energy workers
Chernobyl (Ukraine) 96
Chevron Company 109
chief engineers 31. *See also* electrical engineers
China 160
Citroen 55
civil engineering technicians 114. *See also* petroleum technicians
civil engineers 5, 163. *See also* bioenergy/biofuels workers; wind power engineers
"clean" gas 13
coal gasification engineers vii, **13–21**
Colorado 116, 118
Columbia River Gorge National Scenic Area 70

community affairs representatives 135–136. *See also* renewable energy workers
composite engineers 163. *See also* wind power engineers
Concentrating Solar Power (CSP) technologies 133
Consortium for Energy Efficiency 44
construction engineers 164. *See also* environmental engineers
construction managers and workers 5, 134, 135. *See also* bioenergy/biofuels workers; renewable energy workers
Cool Water plant (Barstow, California) 14
Cornell University 47
Crop Science Society of Agronomy 11
CSP technologies . *See* Concentrating Solar Power technologies
customer service representatives 135. *See also* renewable energy workers

D

Dalton, John 77
Darfur (Western Sudan, Africa) 149
Delphian School (Oregon) 29, 34
demand-side management (DSM) program 37, 44
Denmark 160
Department of Licensing 61
derrick operators 111, 112. *See also* petroleum technicians
Diesel, Rudolf 2, 55
DoE. *See* U.S. Department of Energy
DoL. *See* U.S. Department of Labor
Drake, Edwin 99
drillers 135. *See also* renewable energy workers
drilling engineers 100, 101. *See also* petroleum engineers
drilling technicians 109. *See also* petroleum technicians
drilling workers 135. *See also* renewable energy workers
DSM program. *See* demand-side management program
"dusters" 100

E

Earthworks 127
Eddystone Lighthouse (Devon, England) 47
Edison, Thomas 22, 23, 122, 126
Edison Electric Institute 129
educators 135. *See also* renewable energy workers
Eetrex Company 56
EI. *See* engineer intern
EIA. *See* U.S. Department of Energy, Energy Information Administration
Einstein, Albert 141, 150
EIT. *See* engineer in training
electrical engineers vii, 5, **22–34**, 134, 162, 164. *See also* bioenergy/biofuels workers; renewable energy workers; wind power engineers
electrical technicians 132. *See also* renewable energy workers
electricians 134, 135. *See also* renewable energy workers
Electricity4Gas 61
Electronic Industries Alliance 34
electronics engineers 162. *See also* wind power engineers
embargo 2
Emerson Process Management and Secure Energy Inc. 15
energy conservation technicians vii–viii, **35–45**
energy crisis. *See* fuel crisis/shortages
Energy Policy Act 131
Energy Star appliances 42
Engineering.com 158
engineering managers 100. *See also* petroleum engineers

engineering secretaries 116. *See also* petroleum technicians
engineering technicians 109. *See also* petroleum technicians
engineer intern (EI) 137
engineer in training (EIT) 137
engineers 5, 130, 135. *See also* bioenergy/biofuels workers; renewable energy workers
engine operators 111, 112. *See also* petroleum technicians
EnvironmentalCareer.com 145
environmental engineers 164. *See also* wind power engineers
environmental ethics 119
Environmental Protection Agency "Green Vehicle Guide" 60
environmental scientists 134, 135. *See also* renewable energy workers
Essenes (early sect of Jewish people) 149
Esso Company 109
Evolve-it Motors 56, 62
exploration well 112
Explorer VI (satellite) 150

F

Faraday, Michael 22, 23
farmers 131, 135. *See also* renewable energy workers
Fast Company 29
Federal Energy Regulatory Commission 139
Federal Power Commission 66–67
Fermi, Enrico 77, 87
field service engineers 25. *See also* electrical engineers
fish farmers 134. *See also* renewable energy workers
Ford, Henry 2, 55
foresters 135. *See also* renewable energy workers
4 Times Square (New York City, New York) 151
freelance. *See* self-employment
Frito-Lay 156
Fritts, Charles 150
fuel crisis/shortage 2, 14, 36, 161

G

Gardner, John 154, 156, 158
gasification plant production engineers 16–17. *See also* coal gasification engineers
gaugers 113. *See also* petroleum technicians
GE Energy 166
General Electric Company 23, 82
geochemists 135. *See also* renewable energy workers
geological engineers 100. *See also* petroleum engineers
geological technicians 114. *See also* petroleum technicians
geologists 135. *See also* renewable energy workers
geophysicists 100, 135. *See also* petroleum engineers; renewable energy workers
GeoPrac 51
Georgia Tech 136
geotechnical engineers viii, **46–53**
geotechnics engineers. *See also* geotechnical engineers
Geothermal Education Office 145–146
Geothermal Energy Association 134, 146
Germany 2, 14
Grand Coulee Dam 67
Grandpa's Knob (Castleton, Vermont) 160
Great Wall of China 46
Greek empire 76, 148–149
green car 60
Green Car Magazine 60
The Green Collar Job (Jones) 166
green energy vii. *See also specific types of green energy*
Greener Cars 63
green vehicle designers viii, **54–64**
"Green Vehicle Guide" 60
grid. *See* power grid
ground engineers. *See also* geotechnical engineers
Gulf of Mexico 107, 119
Gullaksen, Bjorn S. 163, 164, 166

Index   175

H

Hahn, Otto 77
Hansen, Peder 162–163, 165–166
Hansen, Pete 56–57, 59, 61, 62
Hawaii 139
Heat Recovery Steam Generator (HRSG) 16
Helios 151
Hoffman Electronics 150
von Holhauzen, Franz 59
Holland 160
Home Depot 151
Home Power 155
How Stuff Works 72
HRSG. *See* Heat Recovery Steam Generator
human resource specialists 131, 135. *See also* renewable energy workers
Hurricane Katrina 49
Hybrid Center 55, 63
hybrid vehicles 55
hydraulic engineers 135. *See also* renewable energy workers
hydroelectric engineers viii, **65–75**
Hydrogasifier 17

I

iCivilEngineer 51
Idaho 139
IEEE. *See* Institute of Electrical and Electronics Engineers
IGCC. *See* Integrated Gasification Combined Cycle
An Inconvenient Truth 60
Indeed.com 20, 127
Industrial Designers Society of America 63
industrial engineers 5. *See also* bioenergy/biofuels workers
information technology workers 135. *See also* renewable energy workers
installers 134. *See also* renewable energy workers
Institute of Electrical and Electronics Engineers (IEEE) 18, 34, 50, 71, 157, 159, 168

Institute of Nuclear Power Operations 91
Institution of Civil Engineers 47, 53
Intech.net (Web site) 155
Integrated Gasification Combined Cycle (IGCC) 14
International Brotherhood of Electrical Workers 126, 129
International Hydropower Association 74–75
International Society for Soil Mechanics and Geotechnical Engineering 53
inventions, modern-day 33
Iowa 139, 161
Iran 110. *See also* Middle Eastern countries
Iraq 110. *See also* Middle Eastern countries
The Italian Job 55

J

JETS. *See* Junior Engineering Technical Society
Jewish Watch International 149
Jobs in Wind Power (Web site) 166
Jones, Van 166
Journal of Petroleum Technology 104
Junior Engineering Technical Society (JETS) 12, 29, 34, 81–82, 85, 103, 108, 121
Junior Solar Sprint 138

K

Key Facts About the Electric Power Industry 129
KoZon Foundation 149
Kuwait 110. *See also* Middle Eastern countries

L

Lara Croft: Tomb Raider 55
lawyers 131. *See also* renewable energy workers
Libby-Owens-Ford Glass Company 150

licensing. *See specific careers*
load dispatchers 125. *See also* power plant workers
loggers 109. *See also* petroleum technicians
Louisiana 104, 116

M

MacCready, Paul 151
machinists 135. *See also* renewable energy workers
Mahar, James R. 70, 74
Maine 139
maintenance technicians 109. *See also* petroleum technicians
Management Information Services vii
marketing professional 135. *See also* renewable energy workers
Marshall Ford Dam 67
Massachusetts 62
mass transit 2
Mattai, Katy 138
mechanical engineers 5, 134, 162, 164. *See also* bioenergy/biofuels workers; renewable energy workers; wind power engineers
mechanical technicians 132. *See also* renewable energy workers
mechanics 131, 135. *See also* renewable energy workers
media relations specialists 135. *See also* renewable energy workers
Mercedes-Benz 55
meteorologists 132. *See also* renewable energy workers
Michigan 62
Michigan Technological University 29–30, 34
Middle Eastern countries 2, 36, 104, 109–110, 160
Midwest Energy Association 137
Midwest Renewable Energy Association 145
Mini (small-engine car) 55
Minnesota 62, 139, 161
Mississippi 118
Mobil Company 109

Model T (car) 2, 55
Montana 15, 139
Morse, Samuel 22
mud loggers 112. *See also* petroleum technicians
mud test technicians 112. *See also* petroleum technicians
Musk, Elon 59

N

NASA 151
National Ambient Air Quality Standards Act 100
National Association of Colleges and Employers 31, 83, 105
National Energy Institute 128
National Hydropower Association 75, 146
National Institute for Certification in Engineering Technologies 39, 44
National Institutes of Health 8
National Mall (Washington, D.C.) 151
National Renewable Energy Laboratory (NREL) 131, 138, 140, 145, 150, 151, 156, 159, 168
   Junior Solar Sprint 138
National Research Council 167
National Society of Professional Engineers 138, 145
The Nature Conservancy (TNC) 119, 121
Netherlands 2
Newsweek 60
New York State 139
Niagara Power Project 66, 72
North American Clean Energy 165
North American Wind Power 165
North Carolina State College 77
Northstar Wind Towers 162
Northwest Youth Corps (Oregon) 40
Norwich University (Northfield, Vermont) 47
NRC. *See* U.S. Nuclear Regulatory Commission

NREL. *See* National Renewable Energy Laboratory
nuclear criticality safety engineers 80. *See also* nuclear engineers
Nuclear Energy Institute 77, 80, 84, 85, 96, 97, 129
nuclear engineers viii, **76–85**
nuclear fuels reclamation engineers 80. *See also* nuclear engineers
nuclear fuels research engineers 80. *See also* nuclear engineers
nuclear health physicists 80. *See also* nuclear engineers
Nuclear Power 2010 program 84, 96
nuclear reactor operators and technicians viii, **86–97**
nuclear reactor operator technicians. *See* nuclear reactor operators and technicians
Nuclear Regulatory Commission. *See* U.S. Nuclear Regulatory Commission

O

Obama, Barack 128, 143
Occupational Outlook Handbook 32
Odeillo, France 150
oil-field equipment test engineers 101. *See also* petroleum engineers
oil-well equipment and services sales engineers 101. *See also* petroleum engineers
oil-well equipment research engineers 101. *See also* petroleum engineers
oil-well fishing-tool technicians 112. *See also* petroleum technicians
Oklahoma 104, 116
OPEC (Organization of Petroleum Exporting Countries) 110
operators 131. *See also* renewable energy workers
Oregon 139
Organization of Petroleum Exporting Countries. *See* OPEC
Oyster Creek, New Jersey 87

P

Pabst, Paul 24, 26–27, 29, 33
Palo Verde, Arizona 87
Parsons (engineering and construction service company) 17
Parthenon (Greece) 46
"passive solar energy" 148, 153
PE. *See* professional engineer
Persia 160
personnel specialists 135. *See also* renewable energy workers
petroleum engineers viii, **98–108**
petroleum technicians viii, **109–121**
Phillips Petroleum 119
photovoltaic (PV) cell 132–133, 134, 143, 151
Plan B (Brown) 166
PMP. *See* certified project manager professional
Popular Mechanics 15–16, 19, 60, 73
power dispatchers 122. *See also* power plant workers
power distributors 122. *See also* power plant workers
power grid 24
power plant operators 122. *See also* power plant workers
power plant workers viii, **122–129**
Pre-Engineering Times (trade publication) 29, 82
process engineers 17. *See also* coal gasification engineers
production engineering technicians 113–114. *See also* petroleum technicians
production engineers 25, 100, 101. *See also* electrical engineers; petroleum engineers
production secretaries 116. *See also* petroleum technicians
production technicians 109. *See also* petroleum technicians
professional engineer (PE) 18, 28, 50, 70, 71, 81, 102, 105, 137, 154, 155, 165
professors 25. *See also* electrical engineers
project managers/directors 17. *See also* coal gasification engineers

public affairs specialists 131, 135–136. *See also* renewable energy workers
pumpers 113. *See also* petroleum technicians
Purdue University 136
PV cell. *See* photovoltaic cell

R

radiation protection technicians 80. *See also* nuclear engineers
R&D Magazine 29
receptionists 135. *See also* renewable energy workers
recreation managers 134. *See also* renewable energy workers
RE&EE. *See* renewable energy and energy efficiency
renewable energy and energy efficiency (RE&EE) vii
renewable energy workers viii, **130–147**
Renewable Energy World 165
Renewable Fuels Association 144
Rensselaer Polytechnic Institute 47, 49
research chief engineers 101. *See also* petroleum engineers
reservoir engineering technicians 114. *See also* petroleum technicians
reservoir engineers 100–101, 135. *See also* petroleum engineers; renewable energy workers
Rhode Island 62
"rigs-to-reefs" program 119
Roman empire 46–47, 141, 148–149
Roosevelt Dam 66
rotary drillers 111–112. *See also* petroleum technicians

S

Salary.com 51, 157, 167
sales engineers 25. *See also* electrical engineers
sales workers 131, 135. *See also* renewable energy workers

San Juan College (New Mexico) 136
Saudi Arabia 110. *See also* Middle Eastern countries
Saussure, Horace-Bénédict de 149
SCADA 69
Science (magazine) 19
Scientific American 73
scientists 4. *See also* bioenergy/biofuels workers; renewable energy workers
secretaries 135. *See also* renewable energy workers
seed production scientists 4. *See also* bioenergy/biofuels workers
self-employment
 bioenergy/biofuels workers 9
 green vehicle designers 61
 petroleum technicians 117
senior operators 86–87. *See also* nuclear reactor operators and technicians
senior reactor operators (SROs) 86–87, 93–94. *See also* nuclear reactor operators and technicians
"seven sisters" 109
Shamrock Island (Corpus Christi, Texas) 119
Sharp Company 150
Shasta Dam 67
Shell Oil Company 109
Shenandoah, Iowa 3
Simply Hired (Web site) 155, 157
"slimhole drilling" 111
Smeaton, John 47
SNG. *See* synthetic natural gas
Society for Sustainable Mobility 63
Society of Petroleum Engineers 104, 105, 106, 108, 121
Society of Women Engineers (SWE) 82–83, 85
soil engineers. *See also* geotechnical engineers
Soil Science Society of Agronomy 11
Solar Building (Albuquerque, New Mexico) 150
Solarbuzz 158, 159
Solar Challenger 151

Solar Cookers International 149
Solar Decathlon competition 151
Solar Electric Power Association 158–159
Solar Energy Industries Association 139, 146, 150, 159
Solar Energy Research Institute. *See* National Renewable Energy Laboratory
solar engineers viii, **148–159**
Solar Living Institute 159
Solar One demonstration project 151
Solar Today 140, 155
South Dakota 139
Southern California Edison 14
SROs. *See* senior reactor operators
Stagg Field (University of Chicago) 87
Stanley, Francis 54
Stanley, Freelan 54
Stanley Steamers 55
Stantec (sustainable design and consulting services) 163, 164
Stottville, New York 66
Strassman, Fritz 77
Stratingh, Sibrandus 54
Supervisory Control and Data Acquisition. *See* SCADA
SWE. *See* Society of Women Engineers
Swedish Road Administration 60
synthetic natural gas (SNG) 17

T

Tarim Basin (China) 104
technicians 131, 134, 135. *See also* renewable energy workers
Tennessee Valley Authority 67
Tesla, Nicholas 23
Tesla Motors 59
testing and commissioning engineers 164. *See also* wind power engineers
Texaco Company 109
Texas 15, 102, 104, 116, 118, 139
Three Mile Island (Harrisburg, Pennsylvania) 96

Time (magazine) 60
Titusville, Pennsylvania 99
TNC. *See* The Nature Conservancy
tool pushers 111, 112. *See also* petroleum technicians
trail planners 134. *See also* renewable energy workers
treaters 113. *See also* petroleum technicians
truck drivers 135. *See also* renewable energy workers

U

UCG. *See* underground coal gasification
underground coal gasification (UCG) 14
Union Electric Company 82
Union of Concerned Scientists 55
unions. *See specific careers*
United Nations 149
Universal Battery, R.M.S. 56
University of California 136
University of Chicago 87
University of Colorado 151
University of Texas at Austin, Petroleum Extension Service 121
University of Wisconsin 136
U.S. Army Corps of Engineers 70, 139
USBR. *See* U.S. Bureau of Reclamation
U.S. Bureau of Mines 14
U.S. Bureau of Reclamation (USBR) 66, 139
U.S. Department of Agriculture 8, 144
U.S. Department of Defense 8
U.S. Department of Energy (DoE)
    Clean Coal Technology Program 14
    Clean Energy Jobs (Web site) 8
    coal gasification engineers 19
    contact information 12, 21, 64, 85
    employment opportunities 82
    Energy Efficiency and Renewable Energy 45, 74, 145

Energy Information
    Administration (EIA) 142, 143
Fossil Energy (Web site) 19
nuclear engineering schools
    listing 81
renewable energy workers 132,
    134, 135
Solar Decathlon competition
    151
U.S. Department of Interior 8
U.S. Department of Labor (DoL)
  earnings data
    bioenergy/biofuels workers 9
    coal gasification engineers 20
    electrical engineers 31
    energy conservation technicians
        41
    green vehicle designers 62
    hydroelectric engineers 73
    nuclear engineers 83
    nuclear reactor operators and
        technicians 94
    petroleum engineers 105
    petroleum technicians 118
    power plant workers 127
    renewable energy workers 141
  employers
    bioenergy/biofuels workers
        7–8
    coal gasification engineers 19
    green vehicle designers 60
    nuclear engineers 82
    solar engineers 156
    wind power engineers 166
  job outlook
    bioenergy/biofuels workers 10
    coal gasification engineers
        20–21
    energy conservation technicians
        41
    geotechnical engineers
    hydroelectric engineers 74
    nuclear engineers 84
    nuclear reactor operators and
        technicians 96
    petroleum engineers 106
    petroleum technicians 120
    power plant workers 128

    solar engineers 158
    wind power engineers 167
  postsecondary training schools
        155, 165
U.S. Department of Transportation
    50–51, 138
U.S. Geological Survey 67, 72
U.S. Nuclear Regulatory
    Commission (NRC) 78–79, 91,
    94, 96, 97, 125
U.S. Reclamation Service. *See* U.S.
    Bureau of Reclamation
Utah 62
Utility Workers Union of America
    126, 129

V

Valmont Company 166
Venezuela 110
Vermont 62
Vespa (motorbike company) 55
Vestas Company 166
Volta, Alexander 22

W

Washington State 139, 161
Water Education Foundation 73, 75
Water Power Act 66
Watts Bar 1 (Tennessee) 87
well loggers 112. *See also* petroleum
    technicians
well-servicing technicians 113. *See
    also* petroleum technicians
Western Electric 150
Westinghouse Company 82
Who Killed the Electric Car? 60
Willamette Falls Lock (Oregon)
    70
Wind Energy Weekly 140
Wind Industry Jobs (Web site) 166
Wind Power (Web site) 165
wind power engineers viii,
    160–168
Windpower Monthly 165
windsmiths 132. *See also* renewable
    energy workers

wind turbine generator (WTG)
  163–164
Wisconsin 62
Worcester Polytechnic Institute
  103, 108
World Exhibition (Paris) 2
World War II 2, 14, 67, 150, 160

WTG. *See* wind turbine generator
Wyoming 15, 116

Y

Your Solar House 150